WITHDRAWN

P9-CAG-789

3 1611 00171 7138

UNIVERSITY
GOVERNORS STATE UNIVERSITY
PARK FOREST SOUTH, ILL.

DATE

F

NOV 22 1983

ENERGY GRAPHICS

PRENTICE-HALL SERIES IN ENERGY

W.J. Kennedy and W.C. Turner, *editors*

Gibson, *Energy Graphics*
Murphy and Soyster, *Economic Behavior of Electric Utilities*

ENERGY GRAPHICS

Graphs, Charts, Tables, Maps
On U.S. and World Production, Consumption, and Reserves

Duncan L. Gibson
Publisher of World Eagle,
The Monthly Social Studies Resource

UNIVERSITY LIBRARY
GOVERNORS STATE UNIVERSITY
PARK FOREST SOUTH, ILL

Prentice-Hall, Inc.
Englewood Cliffs, New Jersey 07632

© 1983 by World Eagle, Inc.
64 Washburn Avenue, Wellesley, Massachusetts 02181

All rights reserved. No part of this book
may be reproduced in any form or by any means
without permission in writing from the publisher.

Library of Congress Catalog Card Number 82-61663

Printed in the United States of America

10   9   8   7   6   5   4   3   2   1

ISBN 0-13-277624-3

Prentice-Hall International, Inc., *London*
Prentice-Hall of Australia Pty. Limited, *Sydney*
Editora Prentice-Hall do Brasil, Ltda., *Rio de Janeiro*
Prentice-Hall Canada Inc., *Toronto*
Prentice-Hall of India Private Limited, *New Delhi*
Prentice-Hall of Japan, Inc., *Tokyo*
Prentice-Hall of Southeast Asia Pte. Ltd., *Singapore*
Whitehall Books Limited, *Wellington, New Zealand*

HD
9502
.A2
G5
1983
c.1

— CONTENTS —

I. U.S. SUMMARY

Sources Of The Energy That The U.S. Is Expected To Consume In 1982 3
1981 U.S. Production Of Energy.
   Also: Quads Of Primary Energy, The U.S., 1981 4
U.S. Energy Flow Diagram, 1981 5
U.S. Energy Production, Consumption, Imports And Exports 6
U.S. Energy Production, Consumption, And Imports 7
U.S. Production Of Energy By Type .... 8 & 9
U.S. Consumption Of Energy By Type .... 10 & 11
U.S. 1981 Energy Production 12
U.S. 1981 Energy Consumption 13
The Role Of Energy Exports And Imports In Total U.S. Merchandise Trade 14
U.S. Energy Supply And Disposition 15
U.S. Consumption Of Energy By End-Use Sector 16
Value Of U.S. Fossil Fuel Production 17
U.S. Production Of Electricity By The Electric Utility Industry
   By Type Of Energy Source 18
The Fuel Sources The U.S. Currently Uses To Generate Electricity 19
U.S. Production Of Electricity By The Electric Utility Industry
   By Type Of Generation 20
U.S. Electric Utility Electricity Flow Diagram, 1981 21
Strategic Petroleum Reserve System Of The U.S. 22
Projected Sources Of U.S. Energy Consumption 23
Energy Required To Produce $100 Of GNP In The U.S. 24
The Future Supply, Sources, And Price Of U.S.
   (And World) Energy Will Depend On Many Variables 25
U.S. Energy Consumption By Sources, 1850-2000 26
Balance Between U.S. Energy Supply And Demand Projections
   By Type Of Energy And Sector, Midprice Case 27
Longer-Term Balance Between U.S. Energy Supply And Demand
   Projections, By Type Of Energy And Sector, Midprice Case 28
Conventional U.S. Energy Supply 29
Several Projections Of U.S. Production In 1990 30

II. WORLD SUMMARY

World Primary Energy Production, 1980 (Preliminary).
   Also: Largest Producers In Each Region 33
Current Consumers Of World Energy 34
World Energy Consumption By Region, 1960 And 1979.
   Also: The Ten Largest Primary Energy Producers, 1980 35
Sources Of World Primary Energy In 1980 (Preliminary) 36
World Primary Energy Production, 1979 37
Percentage Shares Of World Energy Supply, 1970-2000 38
Free World Consumption Of Energy By Region, Total, And Percent
   Of Total, 1979 And Midprice Scenario Projections For 1995 39
Free World Consumption Of Energy By Type 40
Shares Of World Commercial Energy Production And Consumption 41
International Primary Energy Production 42
Energy Consumption Per Capita 43
Japan's Provisional Long-Term Energy Supply And Demand Outlook 44
World Energy Trade, 1975 And 1990 45
West European Energy Imports From U.S.S.R. And World — 1979 46

III. OIL AND GAS

Map: Petroleum Basins Of The World 49
World Oil Production, 1981 50
World Oil Production Off Sharply In 1981: Biggest Year-To-Year
    Decline In History 51
U.S. Petroleum Flow Diagram, 1981 52
U.S. Crude Oil Production 53
U.S. Petroleum Supply And Disposition 54
Crude Oil Production By The World's Major Petroleum
    Exporting Countries, Years 1980 And 1981 55
Estimated Proved Reserves Of Oil As Of January 1, 1982 56
Dependence Of Industrialized Nations On Imported Petroleum, 1980 57
Some Aspects Of Crude Oil Production 58
Where The U.S. Imports Of Oil Came From In 1981 59
Petroleum Imported Directly Into The U.S. From OPEC Countries 60
Map: Major Petroleum Basins, Oilfields, And Gasfields In The U.S.S.R. 61
Map: Major Oil Pipelines In The U.S.S.R. 62
International Production Of Crude Oil 63
International Petroleum Supply And Disposition, 1979 64
Map: International Crude Oil Flow, 1979 65
International Consumption Of Refined Petroleum Products 66
Apparent Consumption Of Oil, Free World, 1950-1979 67
Aspects Of "World Petroleum Availability: 1980-2000" 68
Natural Gas Flow Diagram, The U.S., 1981 69
U.S. Consumption Of Natural Gas By End-User Sector 70
Distribution Of Estimated Proved Reserves Of Oil And Gas
    As Of January 1, 1982 71
Estimated Proved Reserves Of Natural Gas As Of January 1, 1982 72
World Natural Gas Production (Dry), 1980 73
International Supply And Disposition Of Natural Gas, 1979 74
Map: Major Gas Pipelines In The U.S.S.R. 75
Map: International Natural Gas Flow, 1979 76
Map: Europe (Enlarged Inset From Page 76) 77
Western Europe And Soviet Gas 78

IV. COAL

Map: Coalfields Of The United States 81
U.S. Coal Flow Diagram, 1981 82
U.S. Coal Production 83
U.S. Coal Consumption By End-Use Sector 84
Demonstrated Reserve Base Of Coal By Bank, Region, And Potential
    Method Of Mining, January 1, 1979 85
World Coal Production, 1980 86
Estimated International Recoverable Reserves Of Coal, 1979 87
Map: International Coal Flow, 1979 88
U.S. Coal Exports By Country Of Destination 89
History And Projection Of U.S. Coal Exports 90
Map: Coal-Producing Counties In The United States.
    Also: Data On Production And Reserves, Selected Countries 91
Map: Soviet Coal Basins 92
Number Of Federal Coal Acres Under Lease By Business
    Activity Category, 1950-1980 93

V. NUCLEAR

World Commercial Nuclear Power Generation, 1970 And 1980 97
Map: Status Of U.S. Nuclear Powerplants, December 31, 1981 98
Nuclear Power Reactors In Operation And Under Construction
    At The End Of 1981 99
Nuclear 100
Nuclear Electricity Production By Non-Communist Countries 101
World Nuclear Electric Power Production (Net), 1980 102
Installed Nuclear Electric Generating Capacity 103
U.S. Nuclear Powerplant Capacity And Electricity Production 104
U.S. Uranium Resources, January 1, 1981 105
U.S.S.R. 1985 Goal For Electric Generating Capacity 106
The Nuclear Fuel Cycle 107
Fusion Energy 108

VI. NEW AND RENEWABLE

Map: Known And Potential Hydrothermal Resources 111
U.S. Production Of Electricity From Geothermal Sources 112
Illustrative Comparison Of Typical Corn-To-Ethanol/
    Coal-To-Methanol Plants 113
Biomass Resource Base For Alcohol Production, 1990 And 2000. Also:
    Potential Alcohol Yield Of Grains And Potatoes. Also: Estimated U.S.
    Ethanol Production Capacity 114
World Hydroelectric Power Production (Net), 1980 115
Hydropower 116
U.S. Producer Shipments Of Solar Collectors By Type Of Collector
    And Application, 1980 117
Major Potential Solar Power Satellites Environmental Impacts 118-119
World Use Of New And Renewable Sources 120
Map: United States Annual Average Wind Power 121
Typical Annual Energy Output For Small Wind Systems 122
World's First Megawatt-Size Wind Farm 123
New And Renewable Technology Summary-In-Brief 124-127
Glossary 128-132
Acronyms 132

VII. TABLES

Units Of Measure, Conversion Factors, Price Deflators, And Energy
    Equivalents. Also: Approximate U.S. Daily Per Capita
    Consumption Of Types Of Energy In 1980 135-136

I. U.S. SUMMARY

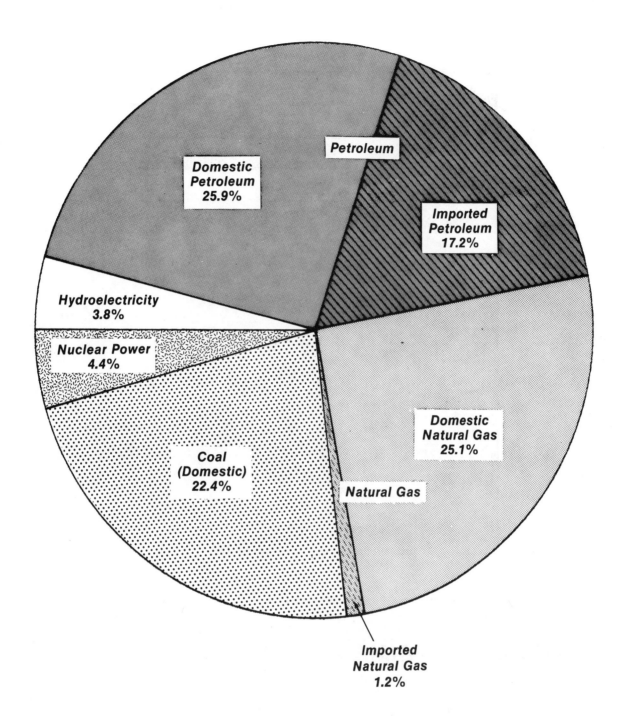

Petroleum

Domestic Petroleum 25.9%

Imported Petroleum 17.2%

Hydroelectricity 3.8%

Nuclear Power 4.4%

Domestic Natural Gas 25.1%

Natural Gas

Coal (Domestic) 22.4%

Imported Natural Gas 1.2%

SOURCES OF THE ENERGY THAT THE U.S. IS EXPECTED TO CONSUME IN 1982

ENERGY GRAPHICS chart based on data from U.S. Department of Energy publication, <u>Short-Term Energy Outlook</u>. August 1981. P.48.

1981 U.S. PRODUCTION OF ENERGY

In Quadrillion (10^{15}) Btu and in Percent

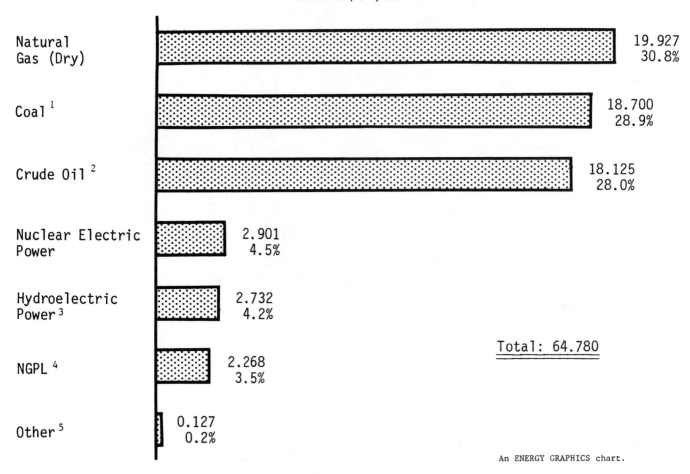

Natural Gas (Dry)	19.927	30.8%
Coal [1]	18.700	28.9%
Crude Oil [2]	18.125	28.0%
Nuclear Electric Power	2.901	4.5%
Hydroelectric Power [3]	2.732	4.2%
NGPL [4]	2.268	3.5%
Other [5]	0.127	0.2%

Total: 64.780

An ENERGY GRAPHICS chart.

1 Includes bituminous coal, lignite, and anthracite.
2 Includes lease condensate.
3 Includes industrial and utility produciton of hydropower.
4 Natural gas plant liquids.
5 Includes geothermal power and electricity produced from wood and waste.

QUADS OF PRIMARY ENERGY, THE U.S., 1981

In Quadrillion (10^{15}) Btu

	81.6	Gross Production from U.S. Sources
+	13.9	Imported (91% of it Oil)
	95.5	
−	17.4	Lost in Converting Energy from One Form to Another and in Transmitting Electricity to End-Users
	78.1	
−	4.3	Exported (68% of it Coal)
	73.8	TOTAL CONSUMED (43% of it Oil)

Data on this page listed in and interpolated from U.S. Department of Energy, 1981 Annual Report to Congress, II, 1. Also: Monthly Energy Review . April 1982.

U.S. ENERGY FLOW DIAGRAM, 1981
(Quadrillion Btu)

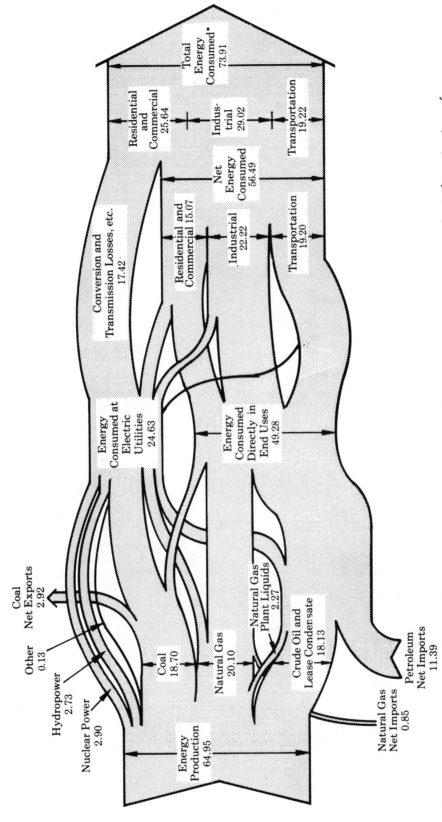

*Total energy consumption with conversion and transmission losses allocated to end use sectors in proportion to the sectors' use of electricity. See footnotes on Table 4.

Note: Changes in stocks, miscellaneous supply and disposition and unaccounted for quantities are not shown.

Source: U.S. Department of Energy. Energy Information Administration. 1981 Annual Report to Congress, II, vii.

U.S. ENERGY PRODUCTION, CONSUMPTION, IMPORTS AND EXPORTS

Quadrillion (10^{15}) Btu

Year	Energy Production[1]	Energy Consumption[2]	Energy Imports[3]	Energy Exports[4]
1973	62.433	74.609	14.732	2.073
1974	61.229	72.759	14.417	2.241
1975	60.059	70.707	14.113	2.389
1976	60.091	74.510	16.838	2.213
1977	60.293	76.330	20.092	2.097
1978	61.231	78.175	19.261	1.952
1979	63.851	78.910	19.620	2.900
1980	65.499	75.913	15.971	3.706
1981	64.780	73.767	13.873	4.336

1. Domestic production of energy includes production of coal (anthracite, bituminous, and lignite), crude oil and lease condensate, natural gas plant liquids, natural gas (dry), electric utility and industrial production of hydropower, and electricity generated from nuclear power, geothermal power, and wood and waste. The volumetric data were converted to approximate heat contents (Btu values) of these energy sources using conversion factors listed in Thermal Conversion Factors.

2. Domestic consumption of energy includes consumption of coal (anthracite, bituminous coal, and lignite), natural gas (dry), refined petroleum products supplied, electric utility and industrial production of hydropower, net imports of electricity produced from hydropower, net imports of coke made from coal, and electricity generated from nuclear power, geothermal power, and wood and waste. Approximate heat contents (Btu values) were derived using conversion factors listed in Thermal Conversion Factors.

3. U.S. energy imports include imports of bituminous coal, crude oil (including crude oil imported for the Strategic Petroleum Reserve), refined petroleum products, natural gas (dry), electricity produced from hydropower, and coke made from coal.

4. U.S. energy exports include bituminous coal and anthracite, crude oil, refined petroleum products, natural gas (dry), electricity produced from hydropower, and coke made from coal.

Source: U.S. Department of Energy. Monthly Energy Review. April 1982.

U.S. ENERGY PRODUCTION, CONSUMPTION, AND IMPORTS

Yearly

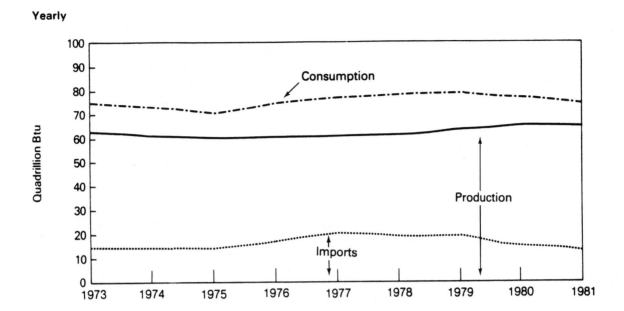

Monthly

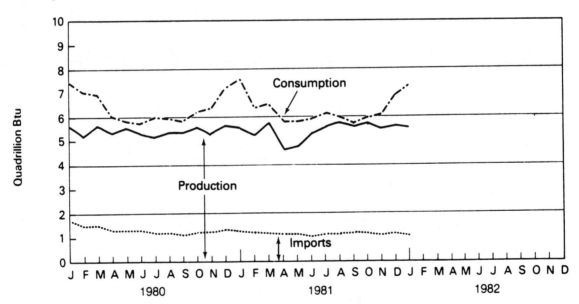

Source: U.S. Department of Energy. Monthly Energy Review. April 1982. P.3.

U.S. PRODUCTION OF ENERGY BY TYPE

Quadrillion (10^{15}) Btu

Year	Coal[1]	Crude Oil[2]	NGPL[3]	Natural Gas (Dry)	Hydro-electric Power[4]	Nuclear Electric Power	Other[5]	Total Energy Produced
1973	14.366	19.493	2.569	22.187	2.861	0.910	0.046	62.433
1974	14.468	18.575	2.471	21.210	3.177	1.272	0.056	61.229
1975	15.189	17.729	2.374	19.640	3.155	1.900	0.072	60.059
1976	15.853	17.262	2.327	19.480	2.976	2.111	0.081	60.091
1977	15.829	17.454	2.327	19.565	2.333	2.702	0.082	60.293
1978	15.037	18.434	2.245	19.485	2.937	3.024	0.068	61.231
1979	17.651	18.104	2.286	20.076	2.931	2.715	0.089	63.851
1980	19.209	18.249	2.254	20.112	2.890	2.672	0.114	65.499
1981	18.700	18.125	2.268	19.927	2.732	2.901	0.127	64.780

Geographic coverage: the 50 United States and District of Columbia.
Totals may not equal sum of components due to independent rounding and the use of preliminary conversion factors after 1979.
[1]Includes bituminous coal, lignite, and anthracite.
[2]Includes lease condensate.
[3]Natural gas plant liquids.
[4]Includes industrial and utility production of hydropower.
[5]Includes geothermal power and electricity produced from wood and waste.

Source: U.S. Department of Energy. Monthly Energy Review. April 1982.

U.S. PRODUCTION OF ENERGY BY TYPE

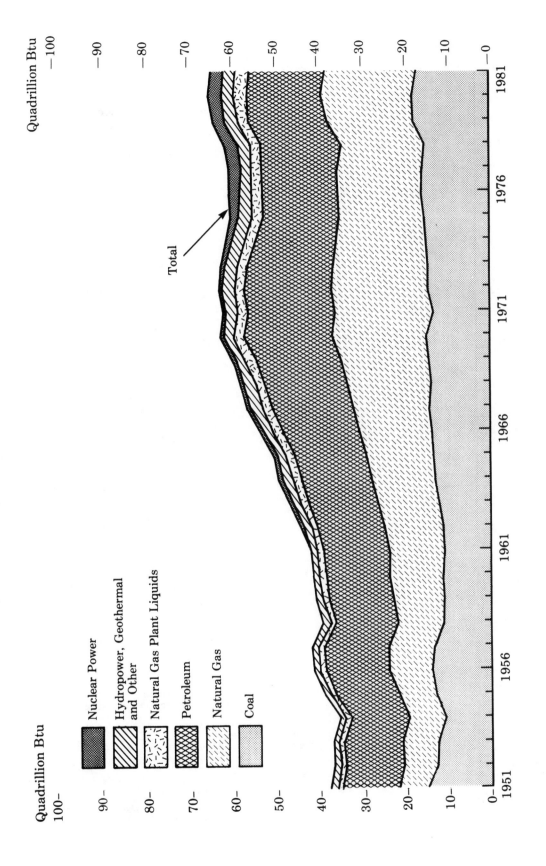

Quadrillion Btu

Nuclear Power

Hydropower, Geothermal and Other

Natural Gas Plant Liquids

Petroleum

Natural Gas

Coal

Total

Quadrillion Btu

Source: U.S. Department of Energy. Energy Information Administration. 1981 Annual Report to Congress, II, 4.

U.S. CONSUMPTION OF ENERGY BY TYPE

Quadrillion (10^{15}) Btu

Year	Coal[1]	Natural Gas (Dry)	Petro-leum	Hydro-electric Power[2]	Nuclear Electric Power	Net Imports of Coal Coke[3]	Other[4]	Total Energy Consumed
1973	13.300	22.512	34.840	3.010	0.910	(0.008)	0.046	74.609
1974	12.876	21.732	33.455	3.309	1.272	0.059	0.056	72.759
1975	12.823	19.948	32.731	3.219	1.900	0.014	0.072	70.707
1976	13.733	20.345	35.175	3.066	2.111	0.000	0.081	74.510
1977	13.964	19.931	37.122	2.515	2.702	0.015	0.082	76.332
1978	13.846	20.000	37.965	3.141	3.024	0.131	0.068	78.175
1979	15.109	20.666	37.123	3.141	2.715	0.066	0.089	78.910
1980	15.461	20.394	34.202	3.107	2.672	(0.037)	0.114	75.913
1981	16.024	19.764	31.998	2.970	2.901	(0.017)	0.127	73.767

Geographic coverage: the 50 United States and District of Columbia.
Totals may not equal sum of components due to independent rounding.
[1]Includes bituminous coal, lignite, and anthracite.
[2]Includes industrial and utility production, and net imports of electricity.
[3]Parentheses indicate exports are greater than imports.
[4]Includes geothermal power and electricity produced from wood and waste.

Source: U.S. Department of Energy. Monthly Energy Review. April 1982.

U.S. CONSUMPTION OF ENERGY BY TYPE

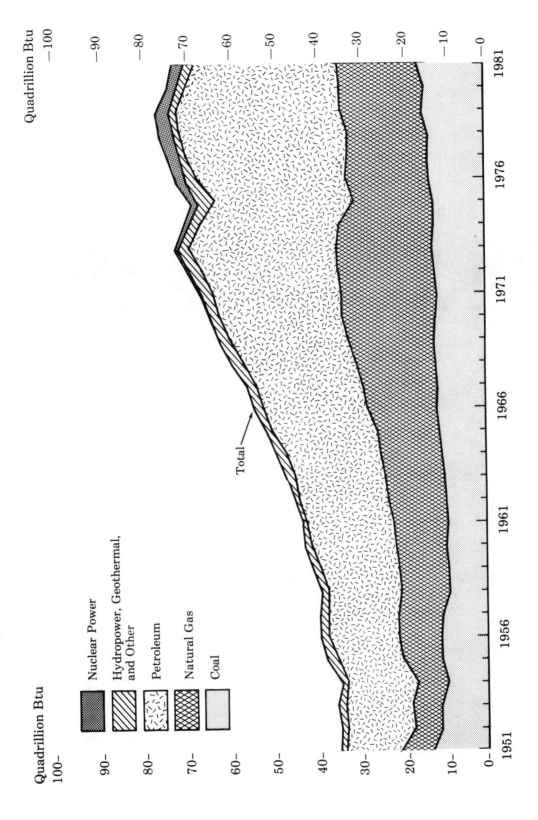

Quadrillion Btu

Nuclear Power

Hydropower, Geothermal, and Other

Petroleum

Natural Gas

Coal

Total

Source: U.S. Department of Energy. Energy Information Administration. 1981 Annual Report to Congress, II, 6.

U.S. 1981 ENERGY PRODUCTION
(in Quadrillion Btu)

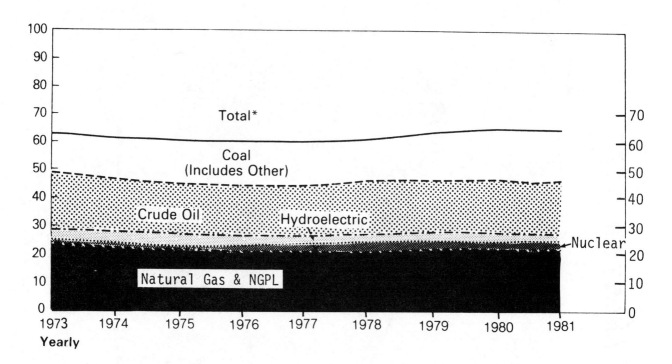

* Btu equivalents for all fuels are cumulated to create total.

U.S. 1981 Production in Quadrillion (10^{15}) Btu

Coal[1] 18.700

Crude Oil[2] 18.125

NGPL[3] 2.268

Natural Gas (Dry) 20.098

Hydroelectric[4] 2.734

Nuclear 2.897

Other[5] 9.127

Total 64.949
versus
Net Consumption
of 73.915

Geographic coverage: the 50 United States and District of Columbia.
Total may not equal sum of components due to independent rounding.
1 Includes bituminous coal, lignite, and anthracite.
2 Includes lease condensate.
3 Natural gas plant liquids.
4 Includes industrial and utility production of hydropower.
5 Includes geothermal power and electricity produced from wood and waste.

Source: U.S. Department of Energy. Monthly Energy Review. March 1982. Pp.4-5.

U.S. 1981 ENERGY CONSUMPTION

(in Quadrillion Btu)

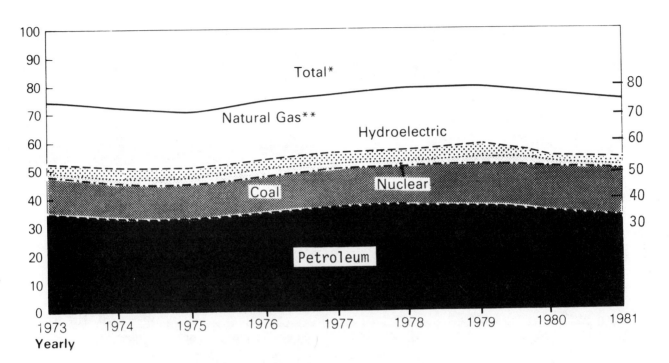

* Btu equivalents for all fuels are cumulated to create total.
** Includes net imports of coal, coke, and other.

U.S. 1981 Consumption in Quadrillion (10^{15}) Btu

Coal[1] 16.011

Natural Gas (Dry) 19.927

Petroleum 31.998

Hydroelectric[2] 2.972

Nuclear 2.897

New Imports of . . . . . (0.017)
    Coal Coke[3]

Other[4] 0.127

Total 73.915
versus
Production
of 64.949

Geographic coverage: the 50 United States and District
of Columbia.
Totals may not equal sum of components due to
independent rounding.
1 Includes bituminous coal, lignite, and anthracite.
2 Includes industrial and utility production, and net
  imports of electricity.
3 Parentheses indicate exports are greater than imports.
4 Includes geothermal power and electricity produced
  from wood and waste.

Source: U.S. Department of Energy. Monthly
Energy Review. March 1982. Pp.6-7.

THE ROLE OF ENERGY EXPORTS AND IMPORTS IN TOTAL U.S. MERCHANDISE TRADE

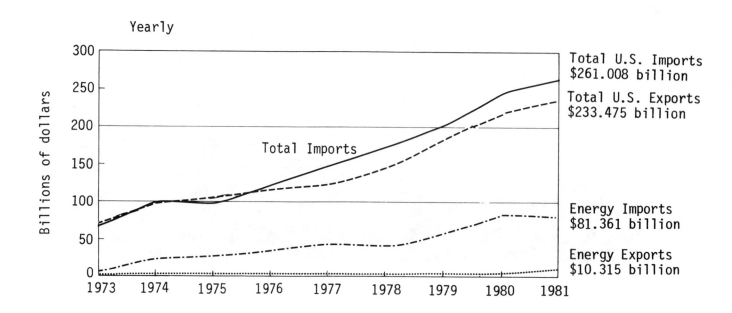

U.S. 1981 Exports and Imports, Billions of Dollars

	Energy	All Other	Total
Exports	$10.315	$223.160	$233.475
Imports	$81.361	$179.647	$261.008
Trade Balance	$-71.046	$+43.511	$-27.535

Source: U.S. Department of Energy. Monthly Energy Review. March 1982. Pp.10-11.

U.S. ENERGY SUPPLY AND DISPOSITION

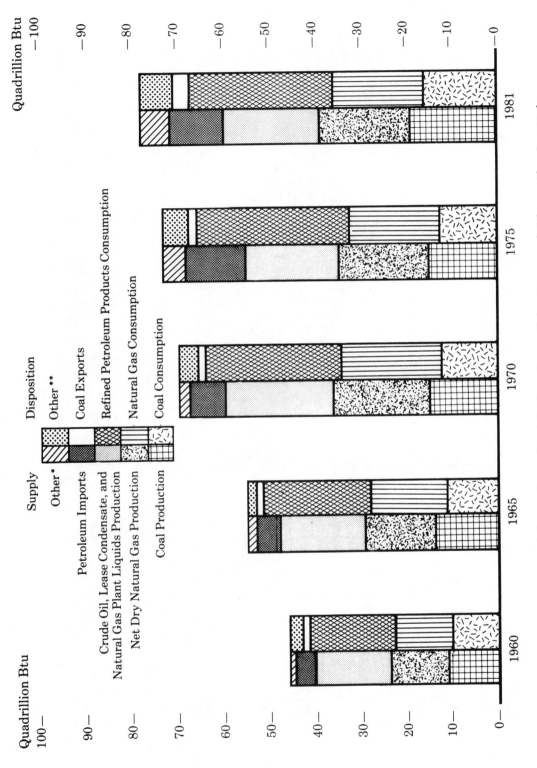

Source: U.S. Department of Energy. Energy Information Administration. 1981 Annual Report to Congress, II, 2.

* Includes: U.S. produced Nuclear Power, Hydro Power, and Other. Also Imported Natural Gas, Other, and Adjustments.
** Includes: U.S. produced Nuclear Power, Hydro Power, and Other. Also Net Imports of Coal Coke and Net Exports of Other.

U.S. CONSUMPTION OF ENERGY BY END-USE SECTOR

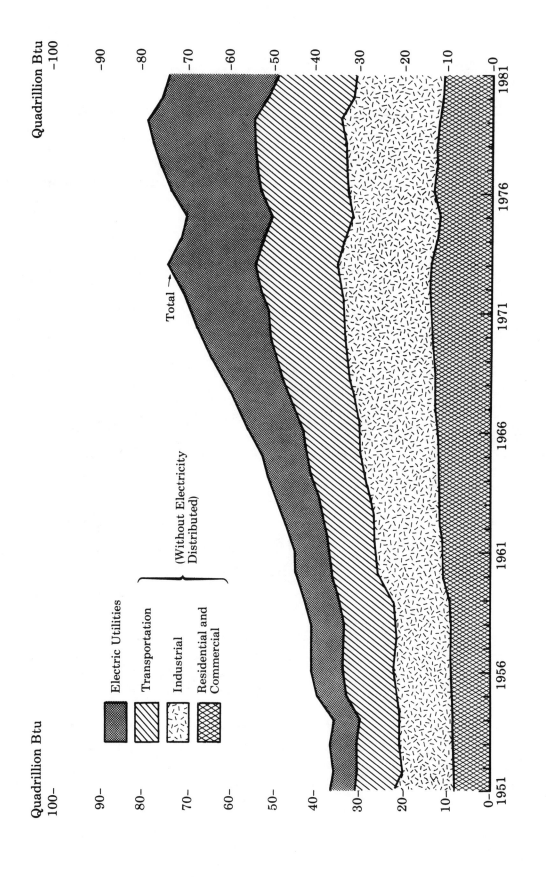

Quadrillion Btu

Electric Utilities

Transportation

Industrial

Residential and Commercial

(Without Electricity Distributed)

Total

Source: U.S. Department of Energy. Energy Information Administration. 1981 Annual Report to Congress, II, 8.

VALUE OF U.S. FOSSIL FUEL PRODUCTION

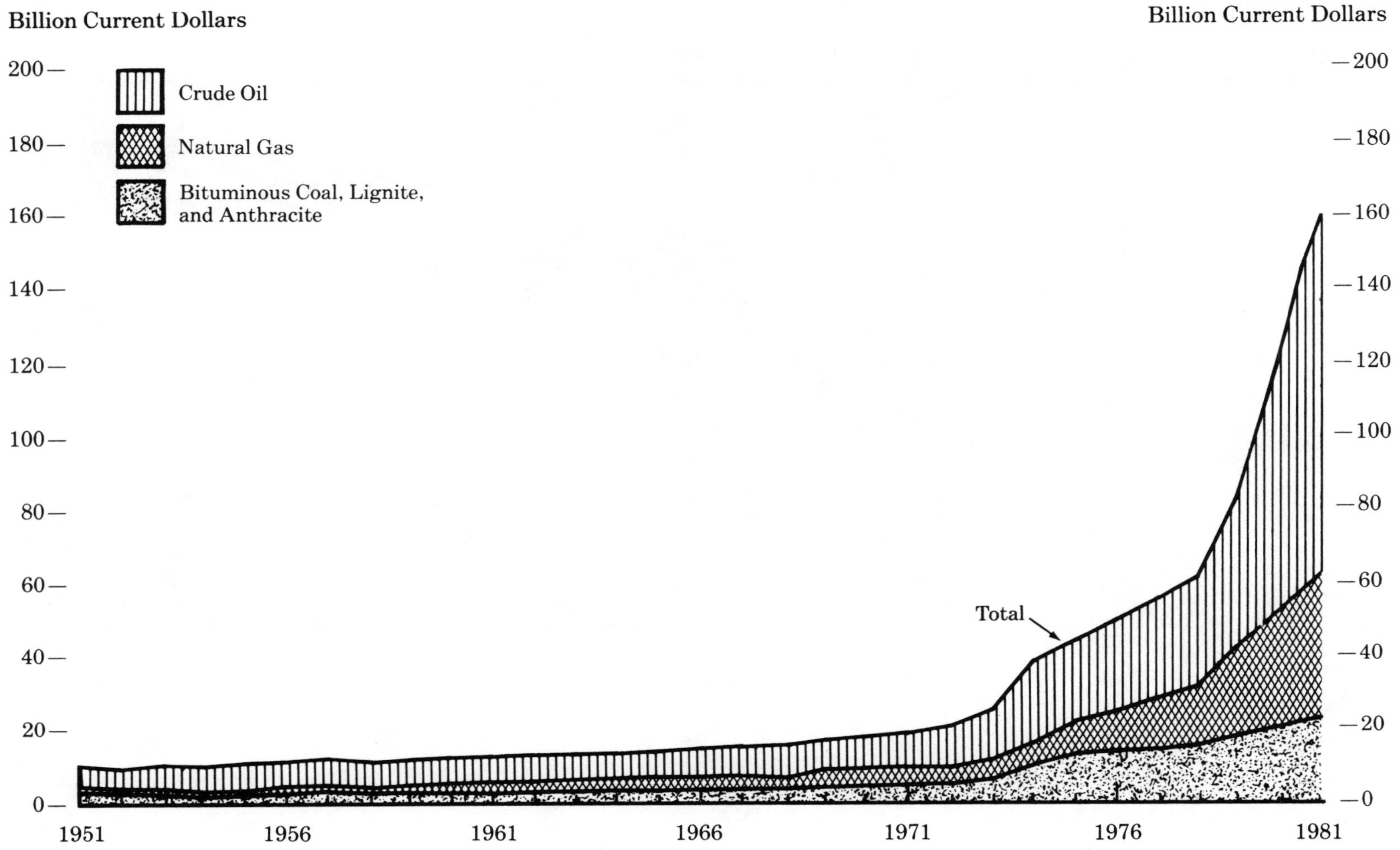

Billion Current Dollars

Crude Oil

Natural Gas

Bituminous Coal, Lignite, and Anthracite

Billion Current Dollars

200 —
180 —
160 —
140 —
120 —
100 —
80 —
60 —
40 —
20 —
0 —

— 200
— 180
— 160
— 140
— 120
— 100
— 80
— 60
— 40
— 20
— 0

Total

1951   1956   1961   1966   1971   1976   1981

Source: U.S. Department of Energy. Energy Information Administration. 1981 Annual Report to Congress, II, 22.

U.S. PRODUCTION OF ELECTRICITY BY THE ELECTRIC UTILITY INDUSTRY BY TYPE OF ENERGY SOURCE

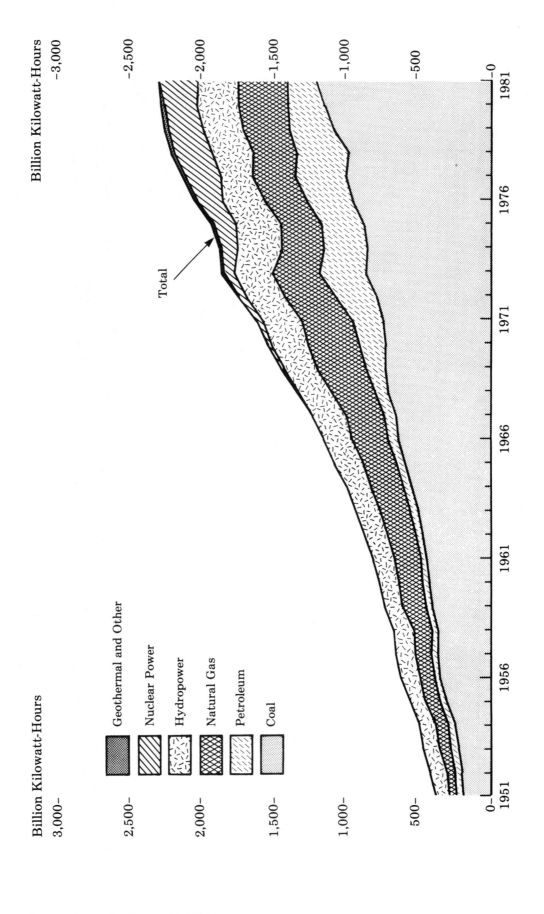

Source: U.S. Department of Energy. Energy Information Administration. 1981 Annual Report to Congress, II, 150.

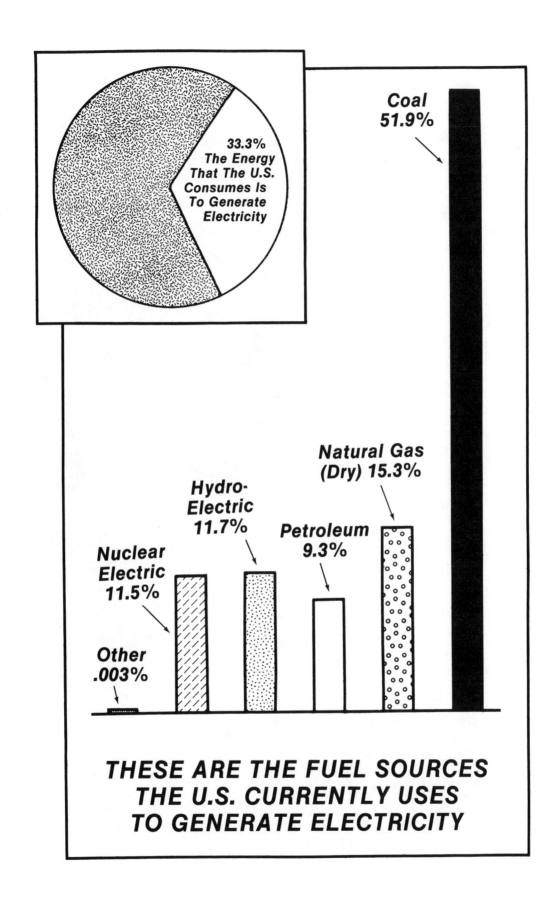

33.3%
The Energy That The U.S. Consumes Is To Generate Electricity

Coal 51.9%

Natural Gas (Dry) 15.3%

Hydro-Electric 11.7%

Petroleum 9.3%

Nuclear Electric 11.5%

Other .003%

THESE ARE THE FUEL SOURCES THE U.S. CURRENTLY USES TO GENERATE ELECTRICITY

ENERGY GRAPHICS chart based on data from U.S. Department of Energy. 1981.

U.S. PRODUCTION OF ELECTRICITY BY THE ELECTRIC UTILITY INDUSTRY BY TYPE OF GENERATION

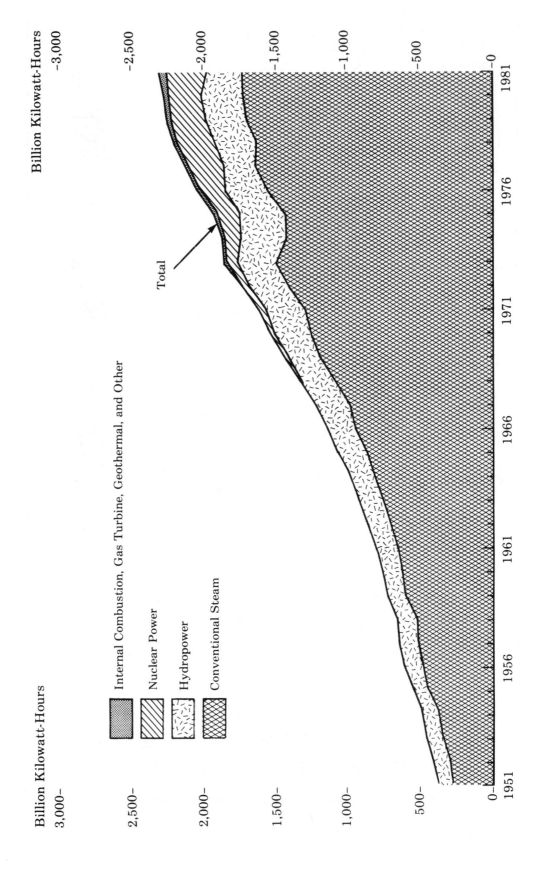

Billion Kilowatt-Hours

Internal Combustion, Gas Turbine, Geothermal, and Other

Nuclear Power

Hydropower

Conventional Steam

Total

Source: U.S. Department of Energy. Energy Information Administration. 1981 Annual Report to Congress, II, 152.

U.S. ELECTRIC UTILITY ELECTRICITY FLOW DIAGRAM, 1981

(Billion Kilowatt-Hours)

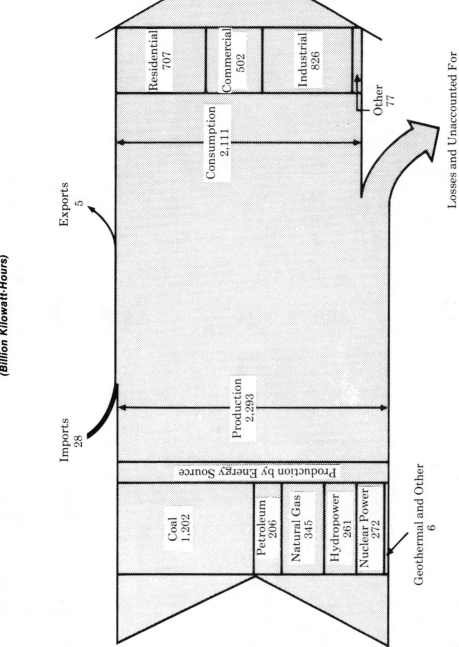

Source: U.S. Department of Energy. Energy Information Administration. 1981 Annual Report to Congress, II, 147.

STRATEGIC PETROLEUM RESERVE SYSTEM OF THE U.S.

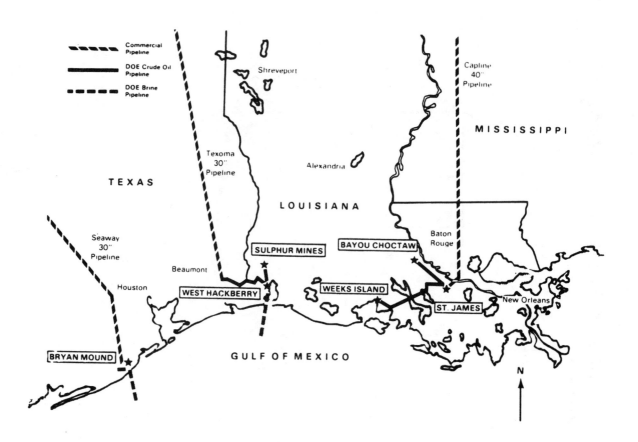

The Energy Policy and Conservation Act of 1975 authorized the creation of a Strategic Petroleum Reserve of up to 1 billion barrels to diminish U.S. vulnerability to a severe petroleum supply interruption. A Strategic Petroleum Reserve Plan became effective in April 1977. Implementation plans have been prepared for the first 750 million barrels of crude oil for storage. Decisions have not yet been made regarding the extent of Government and industry involvement in providing the final 250 million barrels of protection.

The Strategic Petroleum Reserve program is currently being developed in three distinct phases. Phase I consists of the development of 248 million barrels of existing storage capacity in five salt dome sites -- one in Texas and four in Louisiana -- and a marine terminal at St. James, Louisiana, on the Mississippi River (see map above). The Phase I storage sites can be collectively filled at a rate of up to 500,000 barrels per day.

Phase II of the program consists of storage capacity expansion of the Bryan Mound site by 120 million barrels, the West Hackberry site by 160 million barrels, and the Bayou Choctaw site by acquiring an additional

10 million barrels. This 290-million-barrel capacity expansion will be accomplished by creating new storage caverns by means of solution mining (leaching). Construction of Phase II is under way, and leaching at Bryan Mound commenced in March 1980.

During 1980, development continued at the five storage sites and the marine terminal to implement Phases I and II. The storage sites and their planned capacities, plus some 4 million barrels in interconnecting pipelines and surface tanks, will provide 538 million barrels of storage. Phase III of the program entails the creation of 212 million barrels of new storage space.

The Strategic Petroleum Reserve sites have been designed and constructed to supply crude oil to interstate and local pipelines, and to tankers. The Strategic Petroleum Reserve is currently capable of distributing 1.7 million barrels per day through permanent drawdown systems installed at the five operating sites. The drawdown capability will increase to over 3 million barrels per day upon completion of Phase II of the program in 1987.

Source: U.S. Department of Energy. Secretary's Annual Report to Congress. January 1981, I, 16-3 and 4.

PROJECTED SOURCES OF U.S. ENERGY CONSUMPTION

(U.S. Department of Energy, 1980 and 1981)

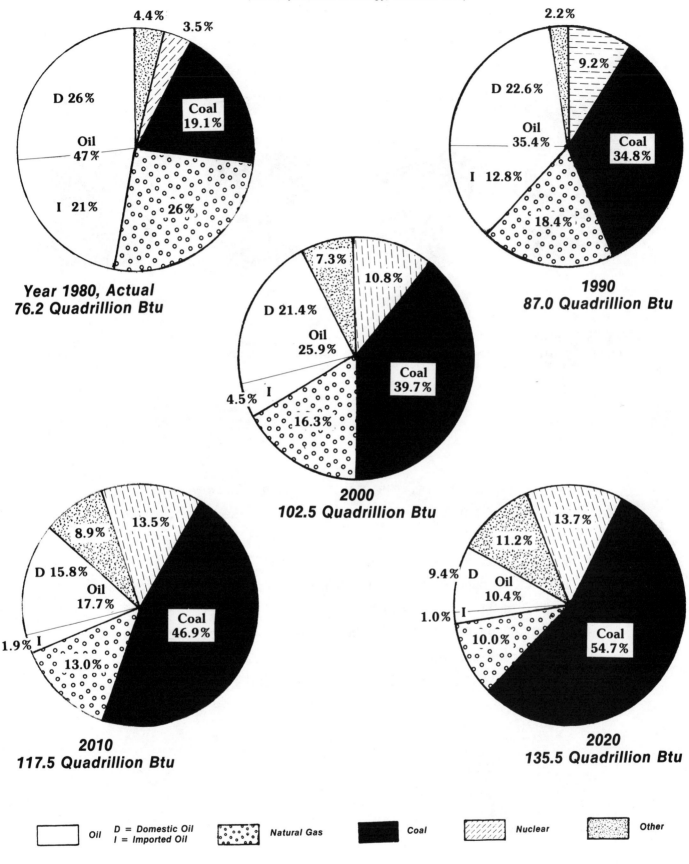

Year 1980, Actual
76.2 Quadrillion Btu

1990
87.0 Quadrillion Btu

2000
102.5 Quadrillion Btu

2010
117.5 Quadrillion Btu

2020
135.5 Quadrillion Btu

Oil   D = Domestic Oil   I = Imported Oil   Natural Gas   Coal   Nuclear   Other

an ENERGY GRAPHICS chart.

ENERGY REQUIRED TO PRODUCE $100 OF GNP IN THE U.S.

(Gallons of oil equivalent per $100 GNP in $1972)

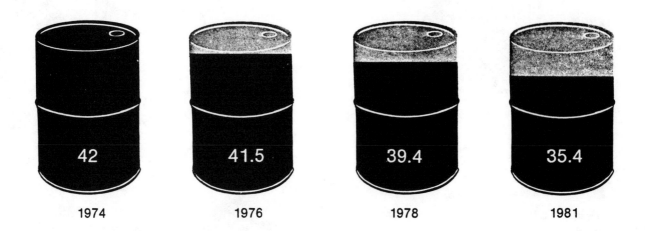

SOURCE: *Monthly Energy Review* (March 1982)

NOTE: The 42-gallon barrel represents the amount of **total energy** required in the production of $100 of GNP in 1974 (GNP is expressed in 1972 dollars to exclude the effects of inflation).

In the United States, the gains in energy conservation during the 1974-81 period came slowly at first. The chart, which uses a 42-gallon barrel of oil as a symbol of total energy consumption, shows that the full "barrel" was needed in 1974 to produce each $100 of gross national product (expressed in 1972 dollars), but that only 35.4 gallons were needed for each $100 of GNP in 1981. Thus, GNP per unit of energy rose 0.7 percent per year from 1974 to 1976, 2.6 percent per year from 1976 to 1978 and 3.6 percent per year from 1978 to 1981.

Source: U.S. Department of Commerce. <u>Business America</u>. May 17, 1982. P.21.

ENERGY TO TURN THE WORLD

Its Future Supply, The Sources, and Its Price
Will Depend on Many Variables

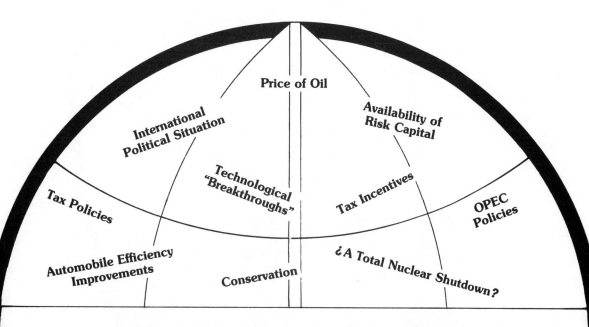

Price of Oil

Availability of Risk Capital

International Political Situation

Technological "Breakthroughs"

Tax Incentives

Tax Policies

OPEC Policies

Automobile Efficiency Improvements

Conservation

¿A Total Nuclear Shutdown?

A "WORLDFULL" OF VARIABLES

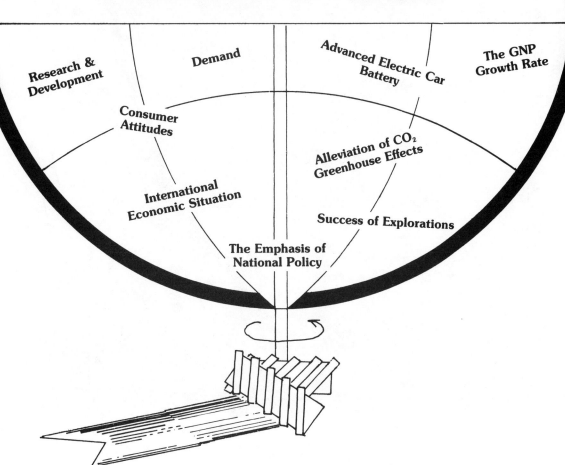

Demand

Advanced Electric Car Battery

The GNP Growth Rate

Research & Development

Consumer Attitudes

Alleviation of CO_2 Greenhouse Effects

International Economic Situation

Success of Explorations

The Emphasis of National Policy

an ENERGY GRAPHCICS chart.

U.S. ENERGY CONSUMPTION BY SOURCES, 1850-2000

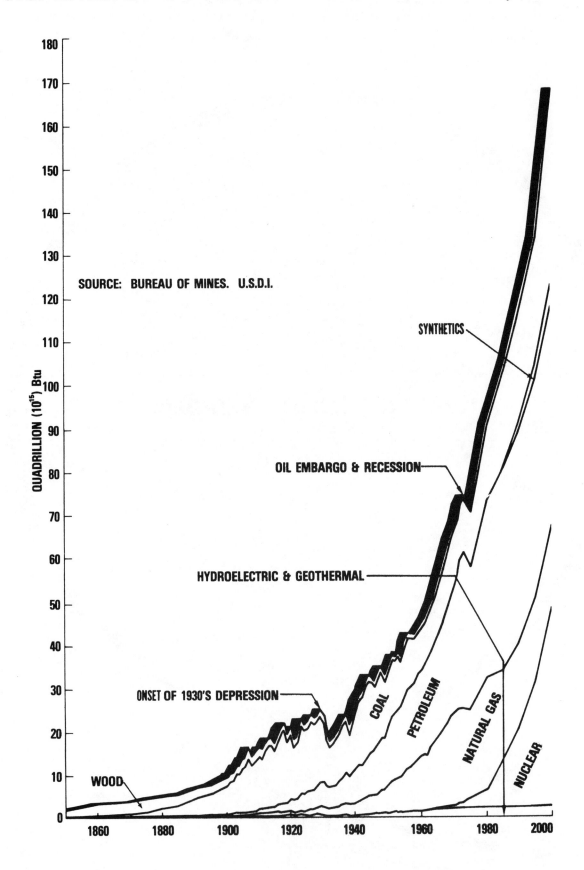

SOURCE: BUREAU OF MINES. U.S.D.I.

SYNTHETICS

OIL EMBARGO & RECESSION

HYDROELECTRIC & GEOTHERMAL

ONSET OF 1930'S DEPRESSION

COAL

PETROLEUM

NATURAL GAS

NUCLEAR

WOOD

QUADRILLION (10^{15}) Btu

Source: U.S. Geologic Survey. Supplemental Reports
to the Second National Water Assessment. 1978.

ENERGY GRAPHICS
64 Washburn Ave., Wellesley, MA 02181

BALANCE BETWEEN U.S. ENERGY SUPPLY AND DEMAND PROJECTIONS BY TYPE OF ENERGY AND SECTOR, MIDPRICE CASE

(Quadrillion Btu per Year)

	History	Projections		
	1980	1985	1990	1995
World Oil Price (1980 dollars per barrel)	33.89	33.00	49.00	67.00
Domestic Energy Supply				
Oil	20.6	19.3	20.0	21.2
Gas	19.8	18.6	18.2	18.7
Coal	18.7	21.9	27.1	33.7
Nuclear	2.7	5.4	7.6	8.6
Other (hydropower, solar, and geothermal)[a]	3.0	3.3	3.5	3.6
Subtotal	64.8	68.5	76.4	85.8
Net Imports				
Oil[b]	13.3	14.7	12.0	10.6
Gas	1.0	0.9	0.7	0.7
Coal	−2.5	−2.7	−3.5	−4.2
Subtotal[c]	12.1	12.8	9.2	7.1
Total Energy Supply	76.8	81.3	85.7	93.0
Domestic Energy Demand				
Residential	9.4	9.0	8.9	9.2
Commercial	6.8	7.0	7.3	8.0
Industrial[d]	23.2	26.4	28.3	30.4
Transportation[e]	19.0	18.9	18.2	18.3
Total End-Use Demand	58.5	61.3	62.7	65.9
Stock Changes, Accounting Errors, and Generating and Transmission Losses	18.3	20.0	23.0	27.0
Total Energy Demand[f]	76.8	81.3	85.7	93.0

[a]Includes gains from electricity generation, synthetics production, and petroleum cracking. Historical data excludes solar.

[b]Figure for 1980 includes imports for additions to the Strategic Petroleum Reserve.

[c]Includes 0.2 quadrillion Btu electricity imported in 1980.

[d]Includes refinery consumption of refined petroleum products and natural gas.

[e]Includes gas transmission losses.

[f]Total supply and consumption estimates include the use of wood to generate electrical power. All other fuel use of wood is at approximately 2 quadrillion Btu.

Sources: Historical data: U.S. Department of Energy, Energy Information Administration, Monthly Energy Review, November 1981, and Natural and Synthetic Gas, 1980 (an Energy Data Report). Estimates of energy end-use consumption were based on the State Energy Data System.

Source: U.S. Department of Energy. Energy Information Administration. 1981 Annual Report to Congress, III, 79.

LONGER-TERM BALANCE BETWEEN U.S. ENERGY SUPPLY AND DEMAND PROJECTIONS, BY TYPE OF ENERGY AND SECTOR, MIDPRICE CASE

(Quadrillion Btu per Year)

	History	Projections		
	1980	1990	2000	2020
World Oil Price (1980 dollars per barrel)	33.90	49.00	75.00	90.00
Domestic Energy Supply[a]				
Oil	21	20	21	17
Natural Gas	20	18	18	15
Coal	19	27	41	77
Nuclear	3	8	11	18
Other (hydropower, solar, geothermal, and biomass)	5	7	11	19
Subtotal[b]	68	80	102	146
Net Imports				
Oil[c]	13	12	8	3
Natural Gas	1	1	1	0
Coal	−3	−4	−6	−6
Subtotal[b d]	11	9	3	−3
Total Energy Supply[b e]	79	89	104	143
Domestic Energy Demand[a]				
Residential	9	10	11	12
Commercial	7	7	9	10
Industrial[f]	23	31	36	47
Transportation[g]	19	18	19	19
Total End-Use Demand[b]	59	66	75	88
Stock Changes, Accounting Errors, and Transmission Losses[h]	20	23	30	55
Total Energy Demand[e]	79	89	104	143

[a]Includes <u>all</u> renewable resources except for 1980, unlike Table 18.
[b]Totals may not add due to rounding.
[c]Figure for 1980 includes imports for additions to the Strategic Petroleum Reserve.
[d]Includes 0.2 quadrillion Btu electricity imported in 1980.
[e]Total supply and consumption estimates include the use of wood to generate electrical power. All other fuel uses of wood are approximately 2.2 quadrillion Btu in 1980 and is explicitly included in consumption and supply for the projections.[a]
[f]Includes refinery consumption of refined petroleum products and natural gas.
[g]Includes gas transmission losses.
[h]Transmission losses are about 9 percent for natural gas and electricity.
Sources: Historical data: U.S. Department of Energy, Energy Information Administration, <u>Monthly Energy Review</u>, November 1981, and <u>Natural and Synthetic Gas</u>, 1980 (an Energy Data Report). Estimates of energy end-use consumption were based on data from the State Energy Data System. Projections were derived using the LEAP model.

Source: U.S. Department of Energy. Energy Information Administration. <u>1981 Annual Report to Congress</u>, III, 104.

CONVENTIONAL U.S. ENERGY SUPPLY

Throughout the 1980-2020 projection period, 75 percent or more of the Nation's primary energy is projected to be derived from conventional domestic sources. These sources include: oil from primary and secondary recovery technologies, natural gas (excluding enhanced recovery), coal, nuclear power, and hydropower. The remainder comes from imports, enhanced recovery of oil and gas, shale oil, and renewable energy sources other than hydropower. The decline of conventional oil and gas production is compensated by the growth of direct coal combustion, synthetic liquids production from coal, and the continuing substitution of electricity generated by coal and nuclear power.

Conventional production of oil and gas is projected to be 32 quadrillion Btu in 2020, down from 36 quadrillion Btu in 1980 and 39 quadrillion Btu in 2000. Achievement of even this production in 2020 depends on significant future discoveries in the frontier areas of Alaska and offshore. Production from the major producing areas of the onshore continental United States is projected to decline more rapidly as the high-quality resources are depleted.

The share of liquid demand projected to be met from conventional oil and gas production decreases throughout the long term. In 2020, conventional oil supplies 8 quadrillion Btu out of the projected 35 quadrillion Btu of consumption, compared to 17 out of 18 quadrillion Btu in 1980. Conventional gas in 2000 supplies 13 quadrillion Btu of the projected 18 quadrillion Btu of consumption compared to 19 out of 20 quadrillion Btu in 1980. The remainder of oil and gas demand is supplied from imports and emerging technologies, including coal conversion. Coal and nuclear power are projected to grow faster than the economy as a whole over the entire 40-year period. From 1980 to 2000, coal use grows at an average annual rate of about 4 percent and nuclear power about 7 percent. In the year 2000, coal is expected to provide two-fifths of primary energy and nuclear one-tenth, compared to the 1980 shares of about one-fourth for coal and one-twenty-fifth for nuclear. The rates of growth from 2000 to 2020 are lower than those of the preceding 20 years because the process of direct substitution of coal and nuclear for oil and gas is largely completed by 2000.

From 2000 to 2020 the domestic coal supply grows at a 3.2-percent average annual rate and nuclear at a 2.5-percent rate. Although nuclear power increases its share of electricity generation at the expense of coal, the total national consumption of coal grows faster because of the demand for synthetic fuels from coal. By 2020, coal supplies over one-half of primary energy and nuclear about one-eighth.

Source: U.S. Department of Energy. Energy Information
Administration. 1981 Annual Report to Congress,
III, 105.

SEVERAL PROJECTIONS OF U.S. PRODUCTION IN 1990

(Quadrillion Btu and 1979 Dollars per Million Btu)

	1978 Actual	1990 Projections						
		1980 *Annual Report*			Other Recent Projections			
		Low	Middle	High	DRI[a]	Bankers Trust[b]	Policy and Evalua-tion[c]	Exxon[d]
World Oil Price (1979 dollars per barrel)	15.50	32.00	41.00	49.00				
Domestic Energy Production								
Oil	20.7	18.8	19.7	20.5	21.4	19.1	17.2	15.7
Gas	19.5	16.1	16.0	15.8	17.0	17.6	17.9	15.7
Coal	15.0	30.3	30.3	30.6	27.2	28.4	26.6	20.9
Nuclear	3.0	8.0	8.0	8.0	7.3	8.5	7.4	7.0
Other	3.0	3.6	3.6	3.6	4.5	5.8	3.6	[e]7.0
Subtotal Domestic Production	61.2	76.8	77.6	78.5	77.4	79.4	[f]75.5	66.3
Oil Imports	17.1	14.1	11.1	8.8	9.4	11.6	14.3	18.8
Gas Imports	0.9	1.1	1.0	1.0	3.1	2.7	2.8	2.1
Coal Imports	-0.9	-2.7	-2.7	-2.7	-2.3	-2.6	-2.1	NA
Subtotal Net Imports	17.2	12.6	9.4	7.1	10.2	11.7	15.0	NA
Total Supply	78.4	89.3	87.0	85.6	87.6	91.1	90.5	87.2
Supply Prices								
Crude Oil (dollars per barrel)								
Domestic Wellhead	9.80	31.66	40.54	48.50	45.75	NA	NA	NA
Imported Landed U.S.	15.86	32.27	41.27	49.27	45.06	NA	NA	NA
Average Refining Acquisition	13.57	32.24	41.21	49.23	45.17	NA	NA	NA
Natural Gas (dollars per million Btu)								
Marginal Price Southwest	2.19	4.20	4.01	3.81	NA	NA	NA	NA
Coal (mine entrance, dollars per ton)								
High Sulfur Bituminous, Northern Appalachia	NA	35.22	34.92	34.92	NA	NA	NA	NA
Low Sulfur Subbituminous, Northwestern Great Plains	NA	9.40	9.40	9.40	NA	NA	NA	NA
Rate of Growth Real GNP 1979–90	NA	2.70	2.60	2.40	3.1	2.4	2.7	[g]2.5

a Data Resources Inc.
b Bankers Trust Company.
c Policy and Evaluation, U.S. Department of Energy, *Reducing U.S. Oil Vulnerability*, November 1980.
d Exxon Company, U.S.A.
e Includes shale oil, coal liquids, coal gases, alcohol, methanol, or gasoline fuels, natural gas and gas fuel liquids from very heavy oils.
f Includes 1.8 quadrillion Btu of biomass used by pulp and paper industry not accounted for by EIA statistics.
g Forecast 1979–2000.
NA Not available

Source: U.S. Department of Energy. 1981.

II. WORLD SUMMARY

WORLD PRIMARY ENERGY PRODUCTION, 1980 (Preliminary)

286.65 quadrillion (10^{15}) Btu

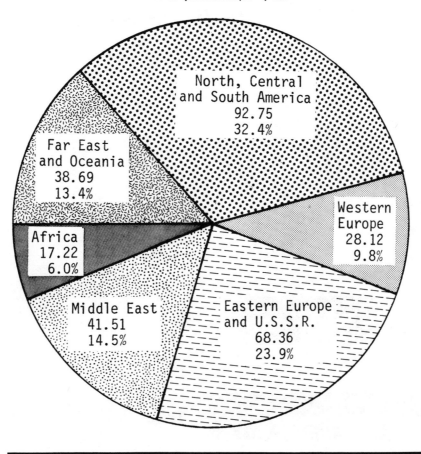

LARGEST PRODUCERS IN EACH REGION

quadrillion (10^{15}) Btu

North, Central, and South America			
U.S.	64.76	Venezuela	5.67
Canada	9.79	Brazil	1.68
Mexico	6.29	Argentina	1.54

Western Europe			
U.K.	8.55	France	2.33
Germany, FRG	4.93	Yugoslavia	1.06
Norway	3.09	Italy	1.06
Netherlands	2.95		

Eastern Europe and U.S.S.R.			
U.S.S.R.	54.45	Romania	2.39
Poland	5.94	Czech.	1.91
Germany, GDR	2.40		

Middle East			
Saudi Arabia	22.09	Iran	3.97
Iraq	5.32	U.A.R.	3.79
Kuwait	4.09		

Africa			
Nigeria	4.41	S. Africa	2.72
Libya	3.98	Egypt	1.45
Algeria	2.86		

Far East and Oceania			
China	20.57	India	2.84
Australia	4.19	Japan	2.30
Indonesia	3.82	N. Korea	1.41

ENERGY GRAPHICS chart and figures based on data from
U.S. Department of Energy. 1980 International Energy
Annual. September 1981.

CURRENT CONSUMERS OF WORLD ENERGY

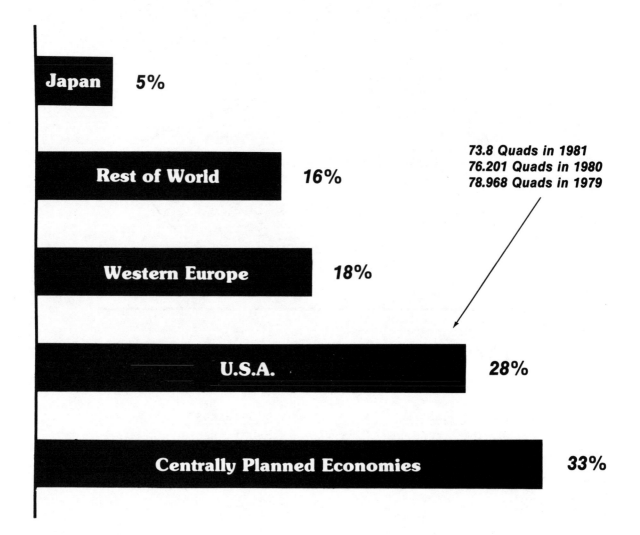

Japan   5%

Rest of World   16%

Western Europe   18%

73.8 Quads in 1981
76.201 Quads in 1980
78.968 Quads in 1979

U.S.A.   28%

Centrally Planned Economies   33%

ENERGY GRAPHICS chart based on data from
U.S. Department of Commerce. 1982.

WORLD ENERGY CONSUMPTION, BY REGION

1960

1979

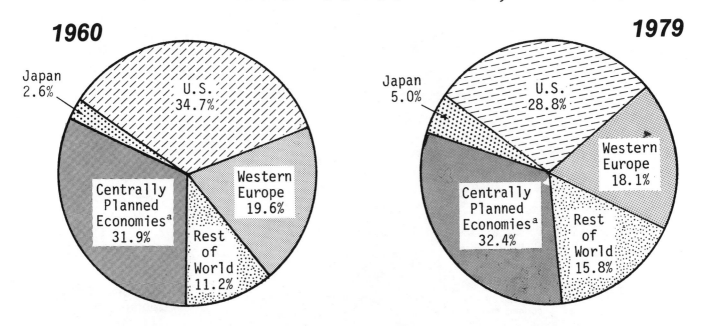

a Includes China, North Korea, Mongolia, Vietnam, Albania, Bulgaria, Czechoslovakia, German Democratic Republic, Hungary, Poland, Romania, and U.S.S.R.

THE TEN LARGEST PRIMARY ENERGY PRODUCERS, 1980

Preliminary Figures

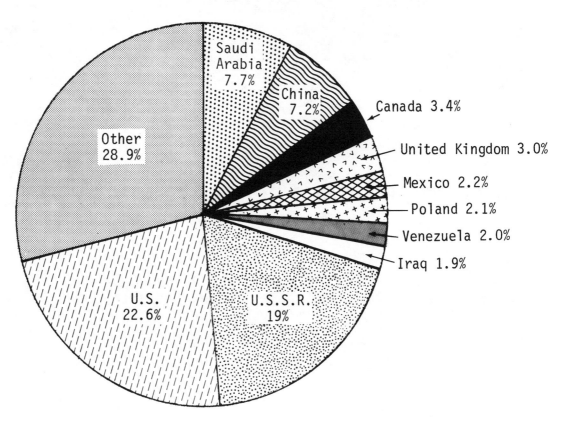

ENERGY GRAPHICS chart based on data in Statistical Abstract of the United States, 1981. Table 1002. P.583. Also on data from U.S. Department of Energy. 1982.

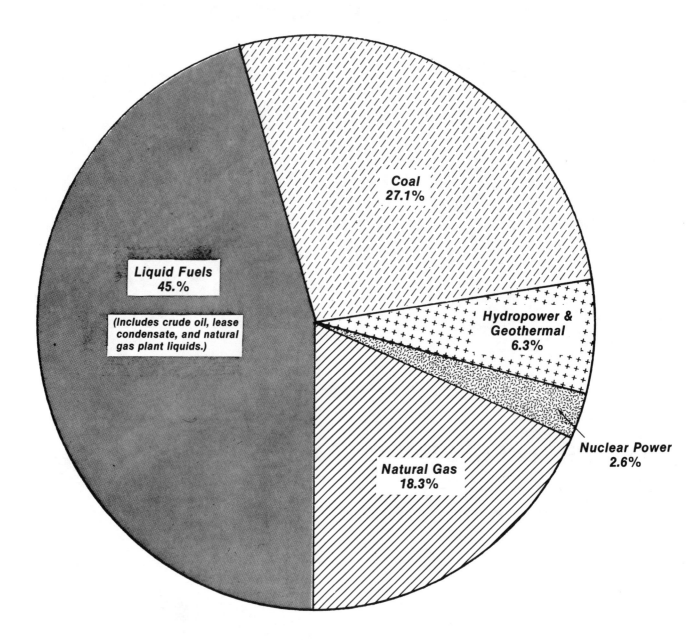

SOURCES OF WORLD PRIMARY ENERGY IN 1980 (Preliminary)

286.65 quadrillion (10^15) Btu

Includes only crude oil, lease condensate, natural gas
plant liquids, dry natural gas, coal, net hydroelectric
power, and net nuclear power.

ENERGY GRAPHICS chart based on data from the U.S. Department
of Energy. 1980 International Energy Annual. September 1981.

WORLD PRIMARY ENERGY PRODUCTION, 1979

Sources, Worldwide
Liquid fuels (crude oil and natural gas liquids). . 47%
Coal. 27%
Natural gas 18%
Hydroelectric power 6%
Nuclear Power 2%

5 Largest Energy Producers,
     in order = 58.2%

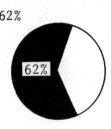

U.S.
U.S.S.R.
Saudi Arabia
China
Canada

5 Largest Crude Oil Producers,
     in order = 58%

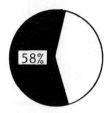

U.S.S.R.
U.S.
Saudi Arabia
Iraq
Iran

2 Largest Dry Natural Gas
     Producers = 62%

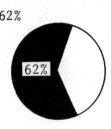

U.S. and U.S.S.R. = 62%
(others,
in order,
the Netherlands,
China, and
Canada)

6 Largest Coal Producers, in
     order, by heat value

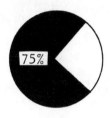

U.S., U.S.S.R.,
and China (57-58%)
Poland 7%,
West Germany 5%,
United Kingdom 5%

5 Largest Hydroelectric
     Producers, in order = 53%

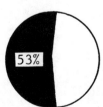

U.S.
Canada
U.S.S.R.
Brazil
Norway

5 Largest Producers of Nuclear
     Power, in order = 73%

U.S.
Japan
U.S.S.R.
United Kingdom
France

ENERGY GRAPHICS charts. Data from U.S. Department of Energy.
1979 International Energy Annual. August 1980.

PERCENTAGE SHARES OF WORLD ENERGY SUPPLY, 1970-2000

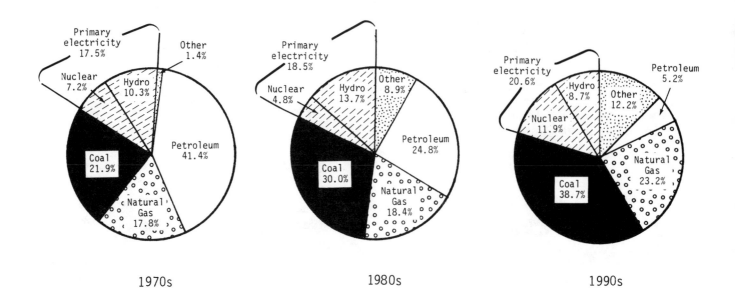

1970s   1980s   1990s

The pattern of energy use prior to the oil price rise in 1973 was unsustainable. When the consumption of oil began to grow faster than additions to reserves, the stage was set for mounting price pressures even in the absence of actions by the oil exporting countries. Although the adjustment to higher prices has not been smooth, their effects have already been marked. In the industrial countries increases in consumption have slowed. But in developing countries continued growth requires rapidly rising quantities of energy.

Adjustments on the supply side will also take place to ease energy bottlenecks in the coming decade. Investments in energy development -- which because of the long lead times made little contribution to adjustment in the past decade -- are now coming to fruition. In the 1980s changes in the composition of supplies are expected to be as important as changes in demand. While oil supplied more than 60 per cent of additional energy in the 1960s, its incre-

mental contribution will be about one quarter in the 1980s. Production of coal is expected to grow twice as fast as that of oil in the 1980s -- coal will gradually replace oil as the world's main source of energy growth. Later, a significant increase in nuclear and synthetic fuels may also be expected.

Until recently, the transition to more expensive energy has been managed relatively smoothly in many of the oil importing developing countries. Although continued progress may be more difficult for some countries in the 1980s, higher energy prices will not, on their own, prevent industrialization and a resumption of faster growth. There will be some changes in comparative advantage, and slower growth in the transition period. Policies to increase domestic energy supplies and to plan for efficient energy use will ease this transition.

ENERGY GRAPHICS charts based on those in The World Bank's World Development Report 1981, as reproduced in Finance & Development, September 1981, p.9. Text from pp. 8 & 9.

FREE WORLD CONSUMPTION OF ENERGY BY REGION, TOTAL, AND PERCENT OF TOTAL, 1979 AND MIDPRICE SCENARIO PROJECTIONS FOR 1995

(Total in Quadrillion Btu)

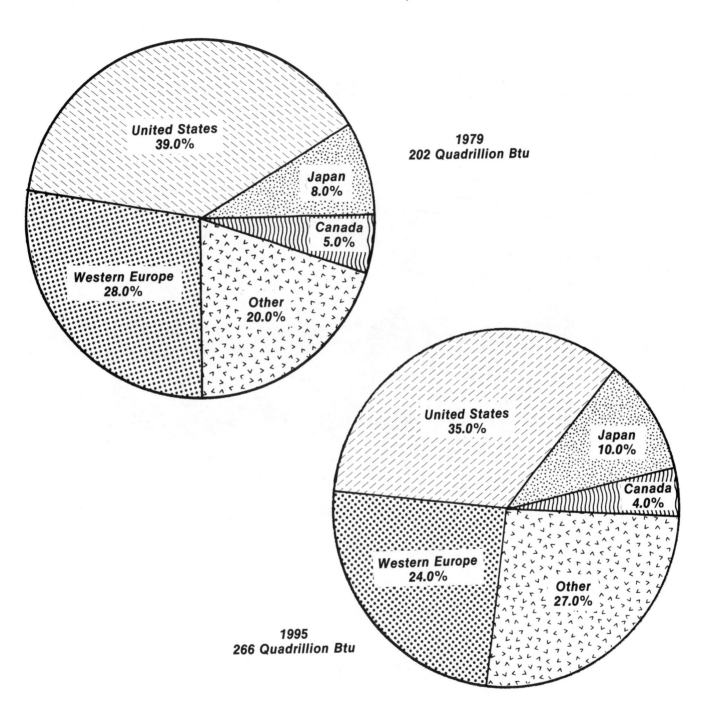

1979
202 Quadrillion Btu

1995
266 Quadrillion Btu

Source: Historical data: U.S. Department of Energy, Energy
Information Administration, 1980 International Energy Annual.

Source: U.S. Department of Energy. Energy Information Administration.
1981 Annual Report to Congress, III, 119.

FREE WORLD CONSUMPTION OF ENERGY BY TYPE

(Projections assume midprice scenario)   *(Quadrillion Btu)*

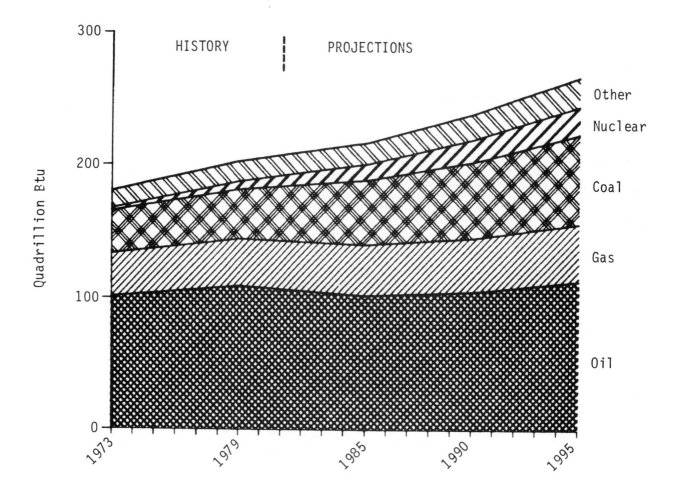

Source: U.S. Department of Energy. Energy Information Administration.
1981 Annual Report to Congress, III, 121.

SHARES OF WORLD COMMERCIAL ENERGY PRODUCTION AND CONSUMPTION

(Percentages)

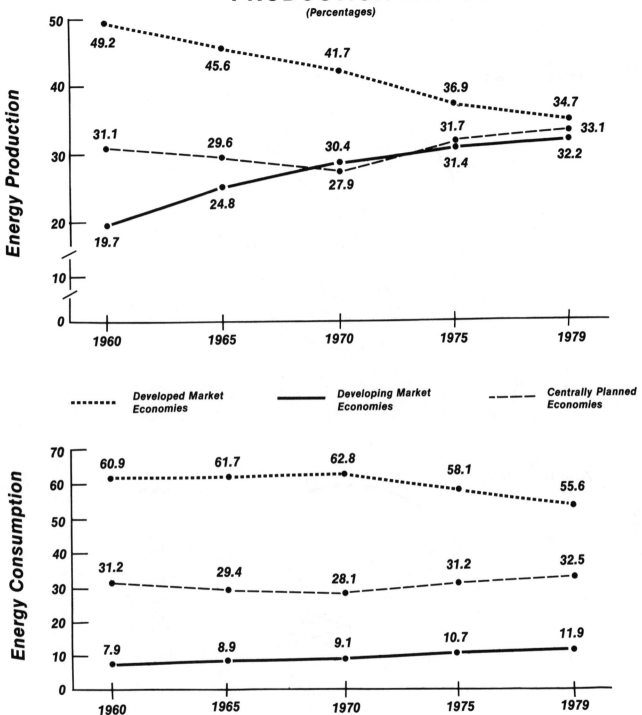

Notes: Commercial energy includes coal, crude petroleum, natural gas liquids, natural gas, and electricity (as generated by hydro, thermal, nuclear, and geothermal processes). This chart does not include non-commercial forms of energy (crop wastes, firewood, etc.), energy which plays a significant role in developing countries.

Data for years other than the five selected have been interpolated by ENERGY GRAPHICS.

ENERGY GRAPHICS charts based on Overseas Development Council's U.S. Foreign Policy and The Third World: Agenda 1982, chart C-6, p.192. That chart was based on data from U.N., Yearbook of World Energy Statistics. 1979.

INTERNATIONAL PRIMARY ENERGY PRODUCTION

(Quadrillion Btu)

1973
World Total: 246.3

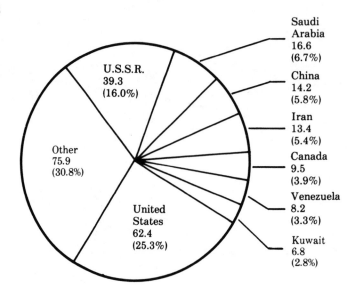

Saudi
Arabia
16.6
(6.7%)

China
14.2
(5.8%)

Iran
13.4
(5.4%)

Canada
9.5
(3.9%)

Venezuela
8.2
(3.3%)

Kuwait
6.8
(2.8%)

U.S.S.R.
39.3
(16.0%)

Other
75.9
(30.8%)

United
States
62.4
(25.3%)

1980
World Total: 286.7

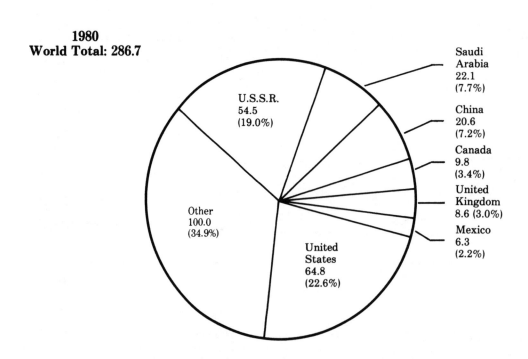

Saudi
Arabia
22.1
(7.7%)

China
20.6
(7.2%)

Canada
9.8
(3.4%)

United
Kingdom
8.6 (3.0%)

Mexico
6.3
(2.2%)

U.S.S.R.
54.5
(19.0%)

Other
100.0
(34.9%)

United
States
64.8
(22.6%)

Source: U.S. Department of Energy. Energy Information Administration.
1981 Annual Report to Congress, II, 18.

ENERGY CONSUMPTION PER CAPITA

(kilos coal equivalent)

	1978	1980		1978	1980
WORLD	2,074	NA	Indonesia	278	220
Argentina. . . .	1,873	1,818	Iran.	1,808	1,246
Australia. . . .	6,622	6,032	Ireland	3,292	2,955
Austria.	4,048	4,160	Israel.	2,362	2,367
Brazil	794	761	Japan	3,825	3,494
Bulgaria	5,020	5,678	Mexico.	1,384	1,770
Canada	9,930	10,241	Netherlands . .	5,327	6,208
Chile.	997	985	Nigeria	106	144
China.	837	602	Philippines . . .	339	316
Cuba	1,168	1,328	Poland.	5,596	5,586
Czechoslovakia .	7,531	6,482	Tanzania.	NA	55
Egypt.	463	496	Turkey.	793	737
France	4,368	4,351	United Kingdom.	5,212	4,942
Germany, FRG . .	6,015	5,727	U.S.A..	11,374	10,410
Hong Kong. . . .	1,657	1,433	U.S.S.R.. . . .	5,500	5,595
India.	178	191			

Sources: 1981 Statistical Abstract, pp.886-887, T.1565 and
1980 Statistical Abstract, pp.916-917, T.1594.

JAPAN'S PROVISIONAL LONG-TERM ENERGY SUPPLY AND DEMAND OUTLOOK

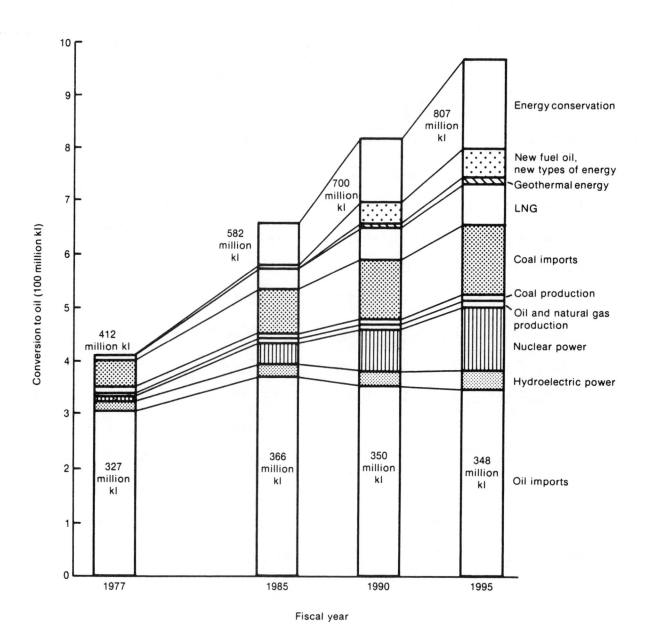

Fiscal year

Note: The projections shown in this diagram are now under revision. According to the long-term oil supply plan published in May 1981, Japan's oil imports during 1985 will total 308 million kl in 1985. This figure includes imports of crude oil and refined products, excluding LNG. The new oil supply plan reflects a reduction in oil imports to Japan from a level of 6.3 - 6.9 million barrels per day for 1985 set at the time of the Tokyo summit in 1979 to a level of about 5.7 million barrels a day for the year 1985. See Tsusho Sangyosho (MITI), *Showa 56-60 Nendo Sekiyu Kyokyu Keikaku* (Oil Supply Plan for 1981-1985), May 27, 1981.

SOURCE: "News from MITI," NR-213 (79-28), Tokyo, Sept. 29, 1979, p. 9.

Source: Congress of the United States. Office of Technology Assessment. <u>Technology & Soviet Energy Availability</u>. November 1981. P. 328.

WORLD ENERGY TRADE, 1975 AND 1990

Japan and Western Europe import 90% of their oil -- more than three-fourths from the Organization of Petroleum Exporting Countries. In 1974, as a result of the 1973 Arab embargoes and OPEC's subsequent price increases, members of the Organization for Economic Cooperation and Development established the International Energy Agency to provide for emergency oil sharing and for collective efforts to develop other sources of energy. Ultimately, the world will have to derive the major portion of its energy from other sources, but there is no immediate substitute for oil.

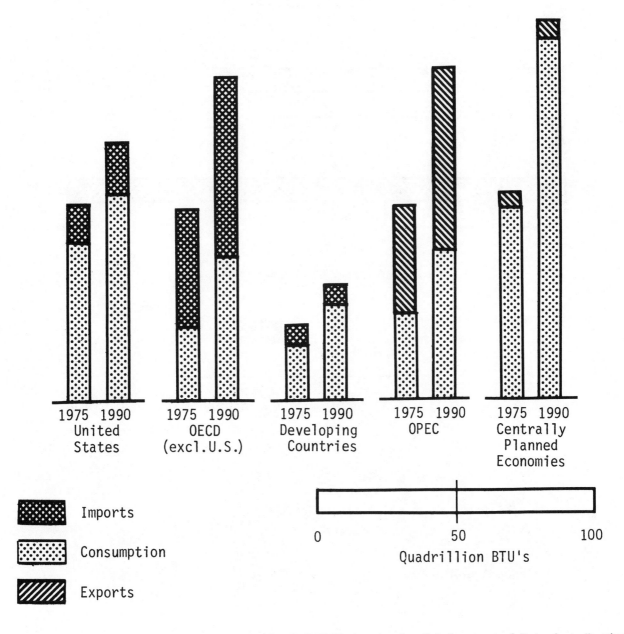

ENERGY GRAPHICS chart based on U.S. Department of State chart, November 1981, which in turn was adapted from Global 2000 Report to the President. Text from U.S. Department of State. Current Policy No. 331. November 1981.

WEST EUROPEAN ENERGY IMPORTS FROM U.S.S.R. AND WORLD — 1979

(million tons of oil equivalent)

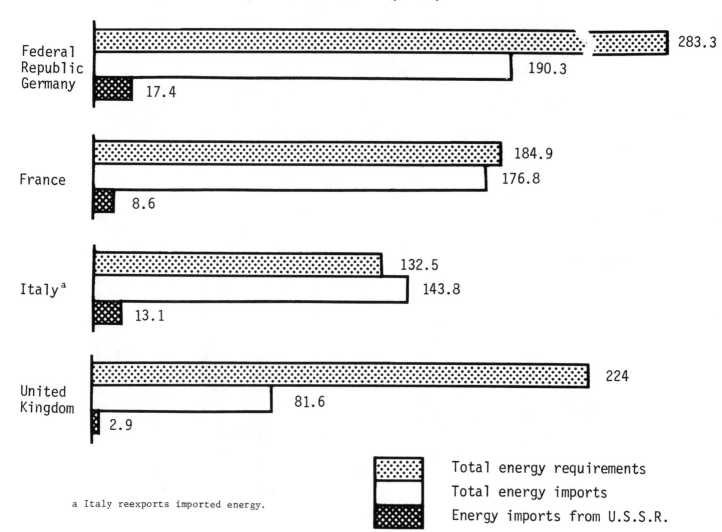

Federal Republic Germany
- 283.3
- 190.3
- 17.4

France
- 184.9
- 176.8
- 8.6

Italy[a]
- 132.5
- 143.8
- 13.1

United Kingdom
- 224
- 81.6
- 2.9

a Italy reexports imported energy.

Total energy requirements
Total energy imports
Energy imports from U.S.S.R.

	Percent imports from U.S.S.R. as part of total	
	Imports	Energy Requirements
Fed. Republic Germany	9.1%	6.1%
France	4.9%	4.7%
Italy	9.1%	9.9%
United Kingdom	3.6%	1.3%

Source: Congress of the United States. Office of
Technology Assessment. Technology & Soviet
Energy Availability. November 1981. P.355.

III. OIL AND GAS

PETROLEUM BASINS OF THE WORLD

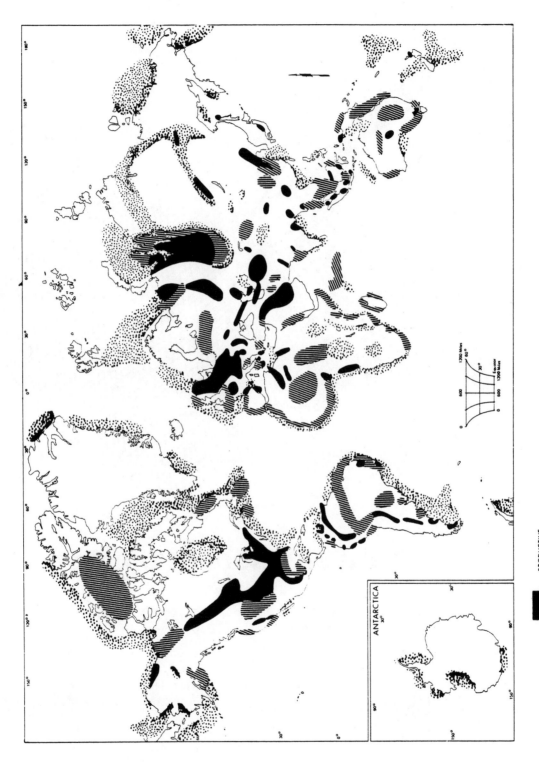

Source: The World Bank.

PRODUCTIVE

PARTIALLY EXPLORED

UNEXPLORED

ANTARCTICA

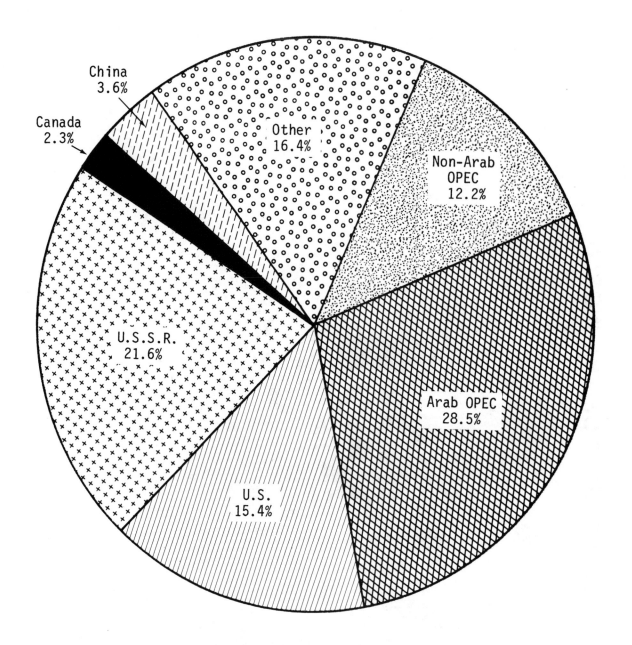

WORLD OIL PRODUCTION, 1981

Arab OPEC		**Non-Arab OPEC**	
percent of world production			
Algeria	1.4%	Indonesia . . .	3.4%
Iraq.	1.8	Iran.	2.4
Kuwait.	2.1	Nigeria	2.6
Libya	2.1	Venezuela . . .	3.8
Qatar	0.7		12.2%
Saudi Arabia. .	17.7		
UAR	2.7		
	28.5%		

ENERGY GRAPHICS chart based on data in U.S. Department of
Energy Monthly Energy Review. April 1982. Pp.99-91. And
in "OGJ Report", Oil and Gas Journal, December 28, 1981.

WORLD OIL PRODUCTION OFF SHARPLY IN 1981
Biggest Year-to-Year Decline in History

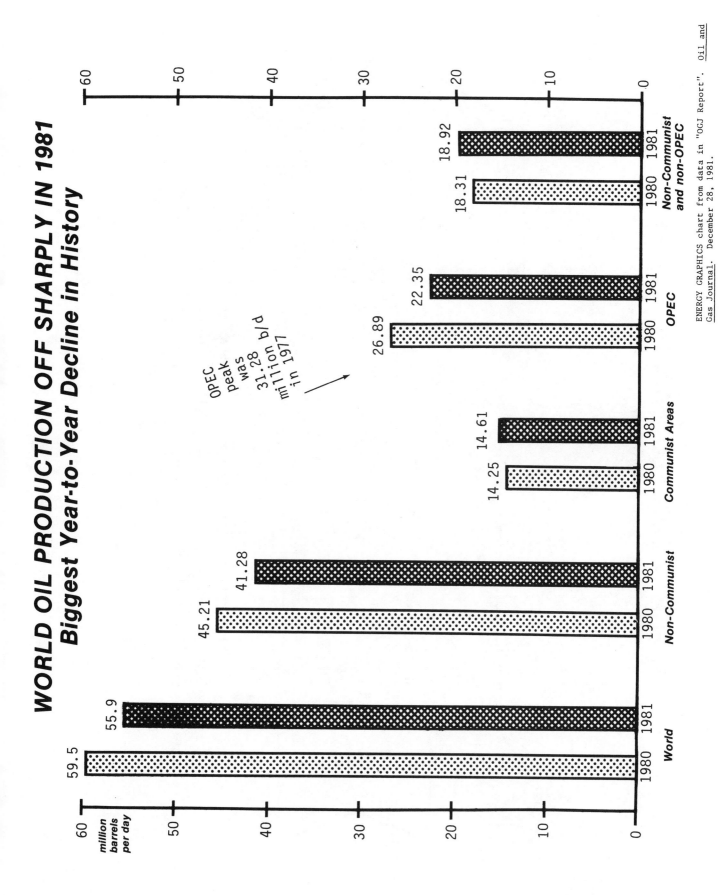

OPEC peak was 31.28 b/d million in 1977

million barrels per day

	World		Non-Communist		Communist Areas		OPEC		Non-Communist and non-OPEC	
1980	59.5		45.21		14.25		26.89		18.31	
1981	55.9		41.28		14.61		22.35		18.92	

ENERGY GRAPHICS chart from data in "OGJ Report". Oil and Gas Journal. December 28, 1981.

U.S. PETROLEUM FLOW DIAGRAM, 1981

(Million Barrels Per Day)

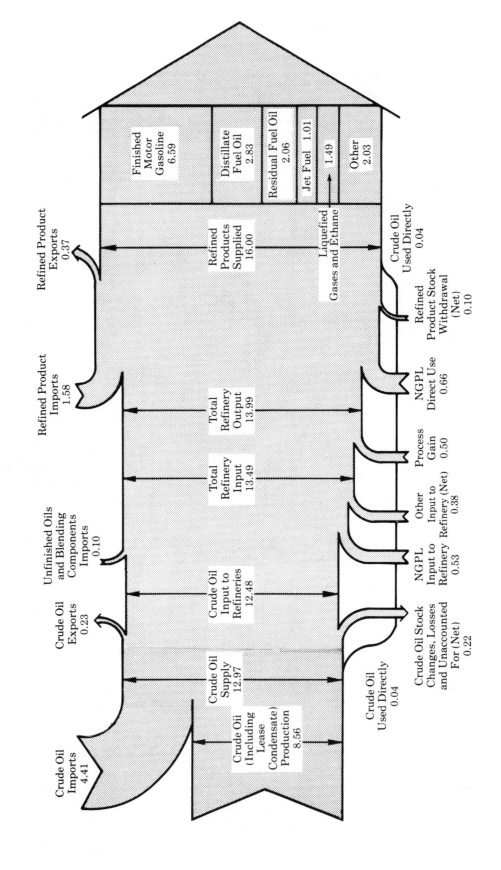

Source: U.S. Department of Energy. Energy Information Administration.
1981 Annual Report to Congress, II, 49.

U.S. CRUDE OIL PRODUCTION

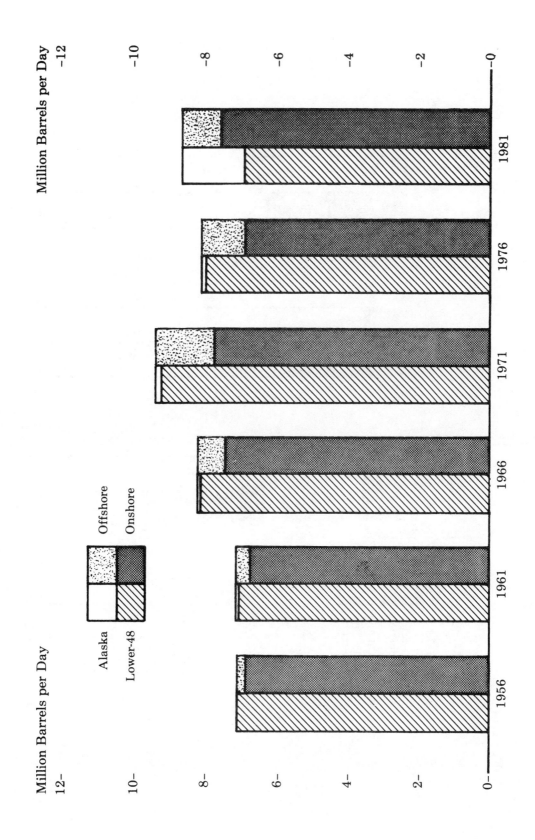

Million Barrels per Day

Million Barrels per Day

Offshore

Onshore

Alaska

Lower-48

12–
10–
8–
6–
4–
2–
0–

1956 1961 1966 1971 1976 1981

-12
-10
-8
-6
-4
-2
-0

Source: U.S. Department of Energy. Energy Information Administration. 1981 Annual Report to Congress, II, 52.

U.S. PETROLEUM SUPPLY AND DISPOSITION

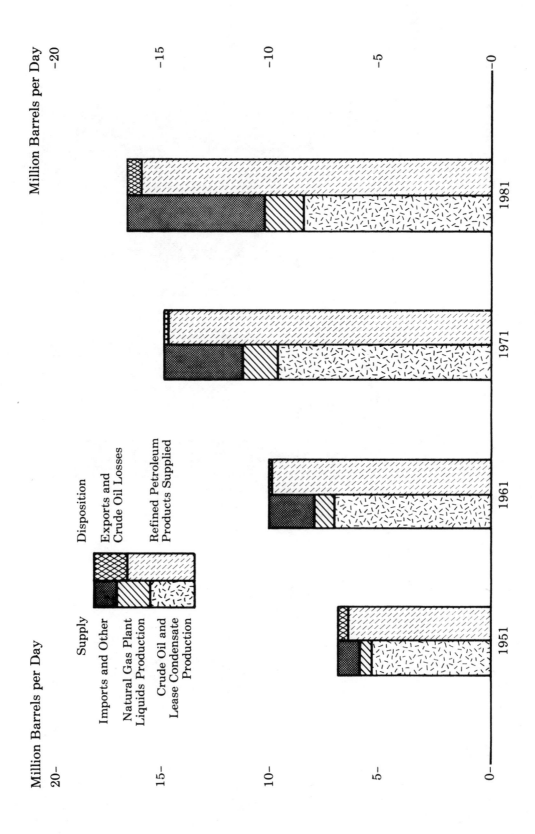

Source: U.S. Department of Energy. Energy Information Administration. 1981 Annual Report to Congress, II, 50.

CRUDE OIL PRODUCTION BY THE WORLD'S MAJOR PETROLEUM EXPORTING COUNTRIES, YEARS 1980 AND 1981

(Thousands of barrels per day, on average.)   *(Percent of World Production.)*

	1980		1981	
Algeria	1,012	1.7%	800	1.4%
Iraq	2,514	4.2%	990	1.8%
Kuwait [1]	1,656	2.8%	1,140	2.1%
Libya	1,787	3.0%	1,145	2.1%
Qatar	472	0.8%	405	0.7%
Saudi Arabia	9,900	16.7%	9,815	17.7%
United Arab Emirates	1,709	2.9%	1,500	2.7%
ARAB OPEC	19,050	32.0%	15,795	28.4%
Indonesia	1,577	2.7%	1,605	3.4%
Iran	1,662	2.8%	1,330	2.4%
Nigeria	2,055	3.5%	1,425	2.6%
Venezuela	2,167	3.6%	2,110	3.8%
TOTAL OPEC [2]	26,890	45.2%	22,630	40.7%
Canada	1,424	2.4%	1,285	2.3%
Mexico	1,937	3.3%	2,310	4.2%
U.K.	1,622	2.7%	1,805	3.2%
United States	8,597	14.5%	8,562	15.4%
China	2,114	3.6%	2,020	3.6%
U.S.S.R.	11,770	19.8%	11,800	21.2%
Other [3]	5,098	8.6%	5,188	9.3%
WORLD	59,452	100.0%	55,600	100.0%

1 Includes about one-half of the former Kuwait-Saudi Arabia Neutral Zone. Production in August 1981 was approximately 257,000 barrels per day.

2 OPEC total includes production in Ecuador and Gabon as well as in the other 11 OPEC countries listed.

3 Other is a calculated total derived from the difference between world production and the nations represented above.

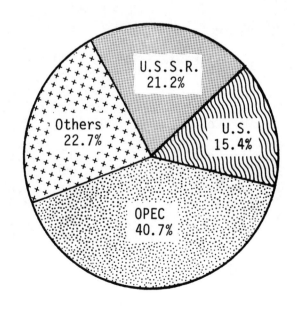

ENERGY GRAPHICS chart based on data from U.S. Department of Energy. Monthly Energy Review. April 1982. Pp.90-91.

ESTIMATED PROVED RESERVES OF OIL
AS OF JANUARY 1, 1982

All reserves figures except those for the U.S.S.R. are proved reserves recoverable with present technology and prices. U.S.S.R. figures are "explored reserves" which include proved, probable, and some possible. U.S. reserves numbers based for the first time on Energy Information Administration estimates which ran some 3 billion bbl. higher than API estimates for 1979, the last year that association provided estimates.

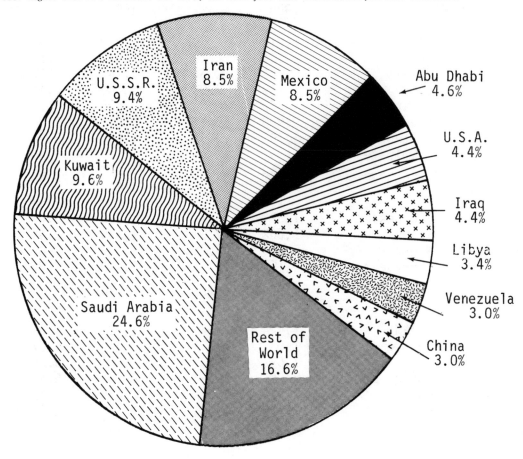

		(1,000 bbl.)						
	WORLD	669,709,150						
1.	Saudi Arabia	164,600,000	24.6%		12.	Nigeria	16,500,000	2.5%
2.	Kuwait	64,480,000	9.6%		13.	United Kingdom	14,800,000	2.2%
3.	U.S.S.R.	63,000,000	9.4%		14.	Indonesia	9,800,000	1.5%
4.	Iran	57,000,000	8.5%		15.	Algeria	8,080,000	1.2%
5.	Mexico	56,990,000	8.5%		16.	Norway	7,620,000	1.1%
6.	Abu Dhabi	30,600,000	4.6%		17.	Canada	7,300,000	1.1%
7.	U.S.A.	29,785,000	4.4%		18.	Divided	6,500,000	1.0%
8.	Iraq	29,700,000	4.4%			(Neutral) Zone		
9.	Libya	22,600,000	3.4%		19.	Qatar	3,434,000	0.5%
10.	Venezuela	20,300,000	3.0%		20.	Egypt	2,930,000	0.4%
11.	China	19,895,000	3.0%		21.	Malaysia	2,800,000	0.4%
					22.	India	2,672,000	0.4%

ENERGY GRAPHICS chart based on data in "OGJ Report". Oil and Gas Journal. December 28, 1981.

DEPENDENCE OF INDUSTRIALIZED NATIONS
ON IMPORTED PETROLEUM, 1980

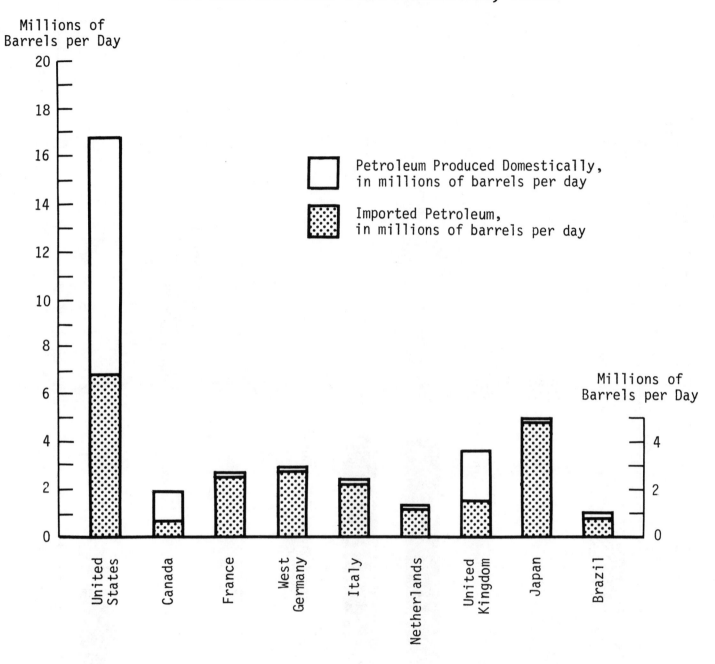

Source: Final Report. U.S. National Alcohol Fuels
Commission. Washington 1981. FUEL ALCOHOL:
An Energy Alternative for the 1980s. P.31.

SOME ASPECTS OF CRUDE OIL PRODUCTION

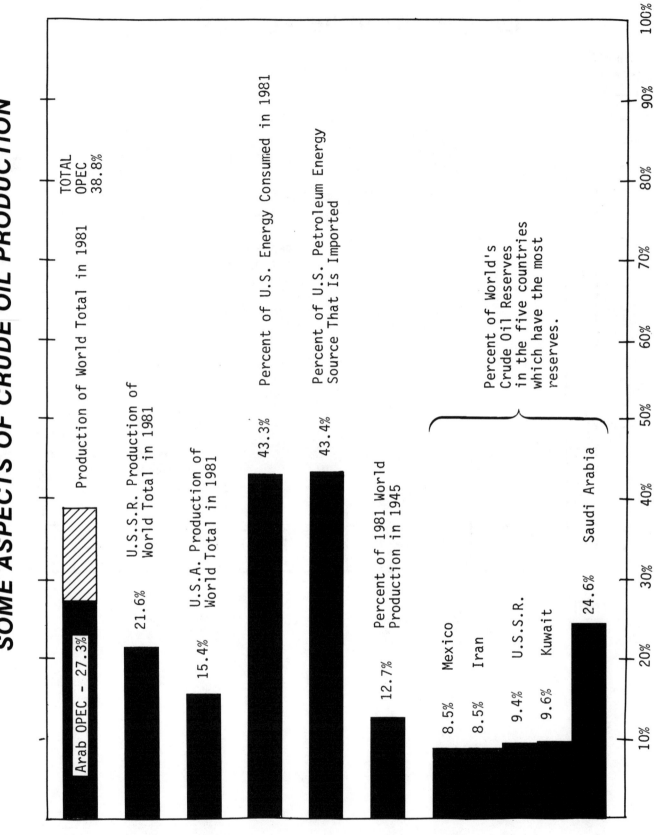

Production of World Total in 1981

TOTAL OPEC 38.8%

Arab OPEC – 27.3%

U.S.S.R. Production of World Total in 1981

21.6%

U.S.A. Production of World Total in 1981

15.4%

Percent of U.S. Energy Consumed in 1981

43.3%

Percent of U.S. Petroleum Energy Source That Is Imported

43.4%

Percent of 1981 World Production in 1945

12.7%

Percent of World's Crude Oil Reserves in the five countries which have the most reserves.

Mexico 8.5%

Iran 8.5%

U.S.S.R. 9.4%

Kuwait 9.6%

Saudi Arabia 24.6%

10% 20% 30% 40% 50% 60% 70% 80% 90% 100%

ENERGY GRAPHICS chart based on data in "OCJ Report". Oil and Gas Journal. December 28, 1981.

WHERE THE U.S. IMPORTS OF OIL CAME FROM IN 1981

43% of the primary energy the U.S. consumed in 1981 was oil.
42% of the oil the U.S. consumed in 1981 was imported.

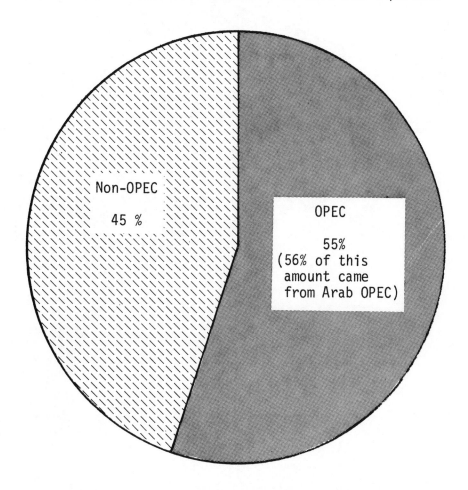

Non-OPEC

45 %

OPEC

55%
(56% of this
amount came
from Arab OPEC)

An Average of 5,981 thousand barrels per day.

Major Sources of U.S. Imported Oil in 1981	Average thousand barrels per day	Percent of U.S. 1981 Imports
Saudi Arabia	1,128	19%
Nigeria.	622	10%
Mexico	523	9%
Canada	445	7%
Venezuela.	404	7%
Indonesia.	364	6%
Virgin Islands	327	5%
Libya	320	5%

ENERGY GRAPHICS chart based on data from U.S. Department of Energy. Monthly Energy Review. March 1982. Pp.34-35.

PETROLEUM IMPORTED DIRECTLY INTO THE U.S. FROM OPEC COUNTRIES

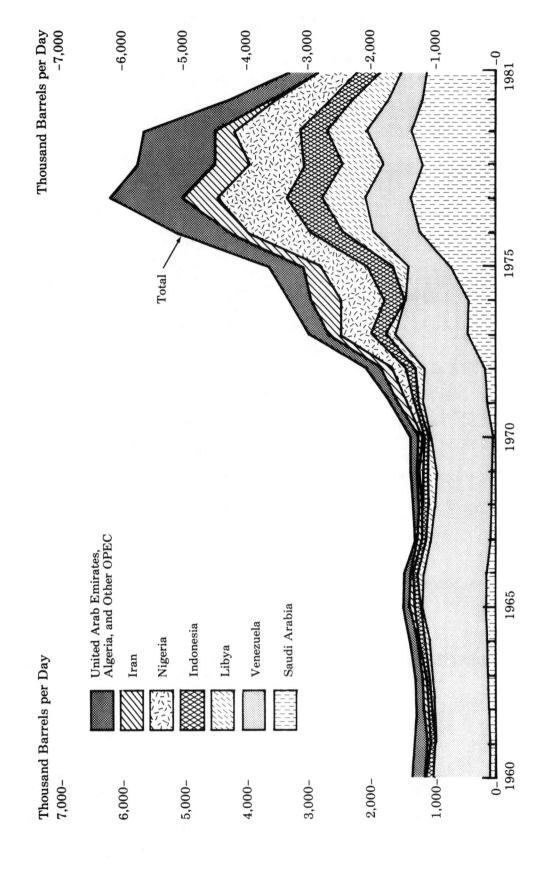

Source: U.S. Department of Energy. Energy Information Administration. 1981 Annual Report to Congress, II, 56.

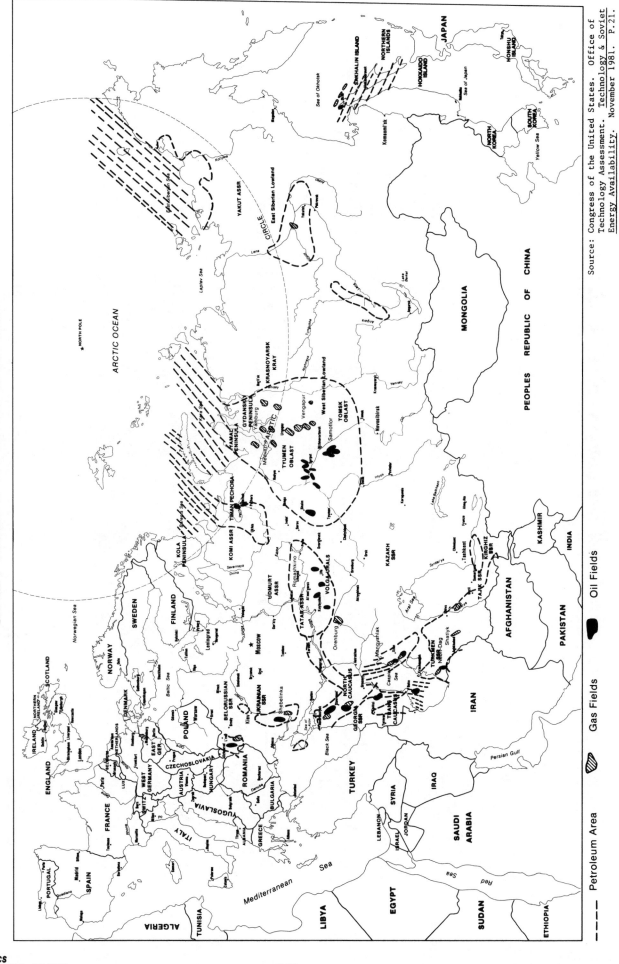

MAJOR PETROLEUM BASINS, OILFIELDS, AND GASFIELDS IN THE U.S.S.R.

Source: Congress of the United States. Office of Technology Assessment. _Technology & Soviet Energy Availability._ November 1981. P. 21.

- - - - Petroleum Area

Gas Fields

Oil Fields

ENERGY GRAPHICS
64 Washburn Ave., Wellesley, MA 02181

MAJOR OIL PIPELINES IN THE U.S.S.R.

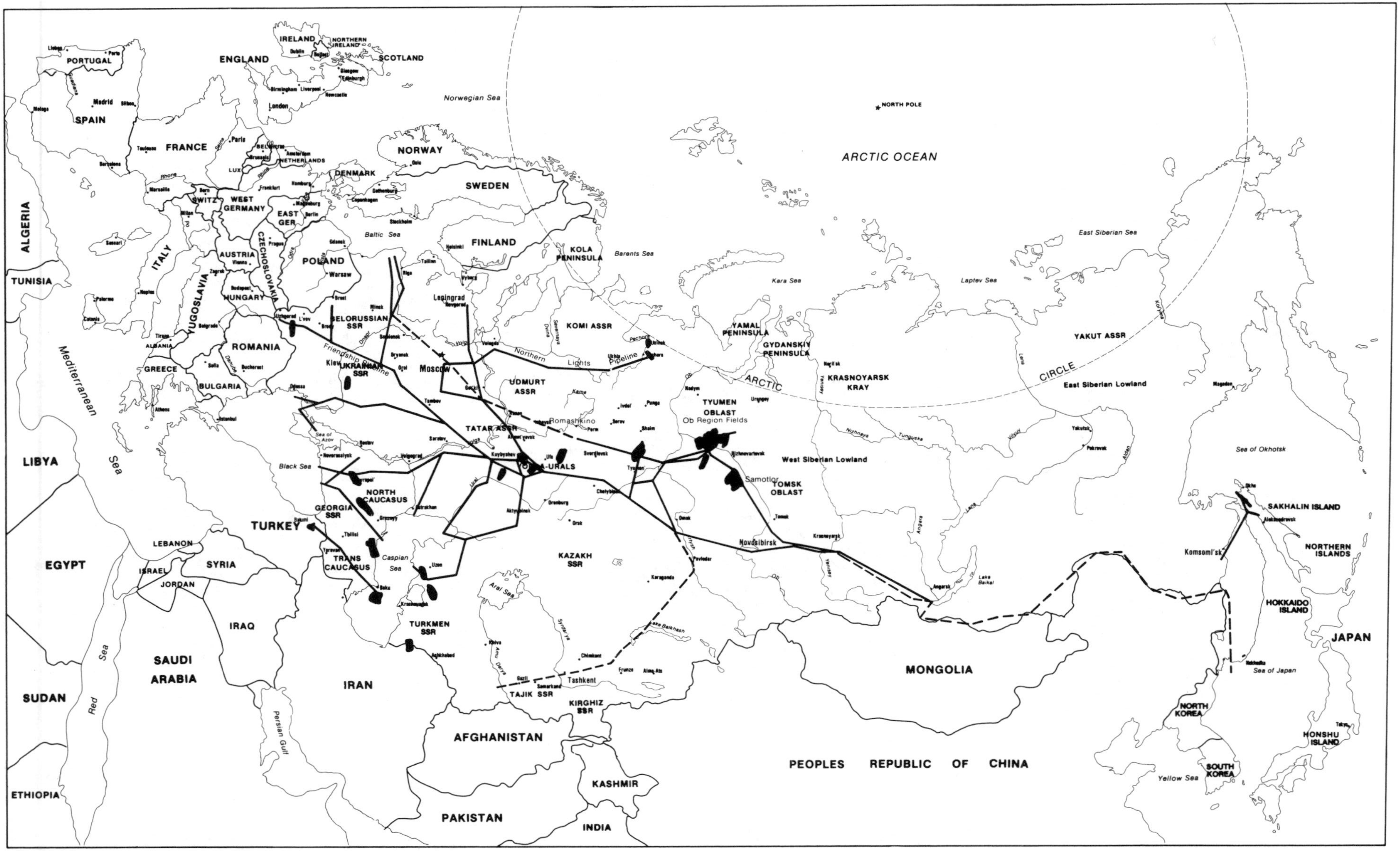

Source: Congress of the United States. Office of Technology Assessment. _Energy & Soviet Energy Availability._ November 1981. P.59.

— Major Pipelines (double and single lines)

- - - Planned

● Oil Fields

Legend / Regions (map labels):

ARCTIC OCEAN · NORTH POLE · ARCTIC CIRCLE · Norwegian Sea · Barents Sea · Kara Sea · Laptev Sea · East Siberian Sea · Baltic Sea · Black Sea · Caspian Sea · Aral Sea · Mediterranean Sea · Red Sea · Persian Gulf · Sea of Okhotsk · Sea of Japan · Yellow Sea

YAMAL PENINSULA · GYDANSKY PENINSULA · KOLA PENINSULA · YAKUT ASSR · KRASNOYARSK KRAY · KOMI ASSR · UDMURT ASSR · TATAR ASSR · TYUMEN OBLAST · Ob Region Fields · OMSK OBLAST · West Siberian Lowland · East Siberian Lowland · URALS · KAZAKH SSR · KIRGHIZ SSR · TAJIK SSR · TURKMEN SSR · BELORUSSIAN SSR · UKRAINE SSR · NORTH CAUCASUS · GEORGIAN SSR · TRANS CAUCASUS · KASHMIR

Moscow · Leningrad · Friendship Pipeline · Kiev · Lvov · Northern Lights · Ukraine Pipeline · Romashkino · Samotlor · Tashkent

ENGLAND · SCOTLAND · IRELAND · NORTHERN IRELAND · FRANCE · SPAIN · PORTUGAL · NORWAY · SWEDEN · FINLAND · DENMARK · NETHERLANDS · WEST GERMANY · EAST GERMANY · SWITZ. · LUX. · POLAND · CZECHOSLOVAKIA · AUSTRIA · HUNGARY · ROMANIA · YUGOSLAVIA · ITALY · GREECE · ALBANIA · BULGARIA

MONGOLIA · PEOPLES REPUBLIC OF CHINA · JAPAN · NORTH KOREA · SOUTH KOREA · SAKHALIN ISLAND · NORTHERN ISLANDS · HOKKAIDO ISLAND · HONSHU ISLAND

TURKEY · SYRIA · LEBANON · ISRAEL · JORDAN · IRAQ · IRAN · SAUDI ARABIA · AFGHANISTAN · PAKISTAN · INDIA · EGYPT · LIBYA · TUNISIA · ALGERIA · SUDAN · ETHIOPIA

INTERNATIONAL PRODUCTION OF CRUDE OIL

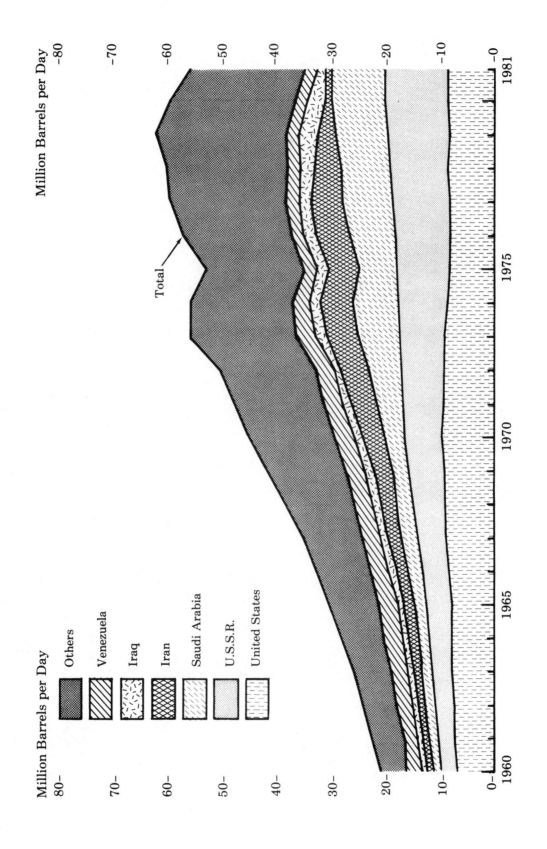

Million Barrels per Day

Million Barrels per Day

Others
Venezuela
Iraq
Iran
Saudi Arabia
U.S.S.R.
United States

Total

1960 1965 1970 1975 1981

Source: U.S. Department of Energy. Energy Information Administration.
1981 Annual Report to Congress, II, 78.

INTERNATIONAL PETROLEUM SUPPLY AND DISPOSITION, 1979

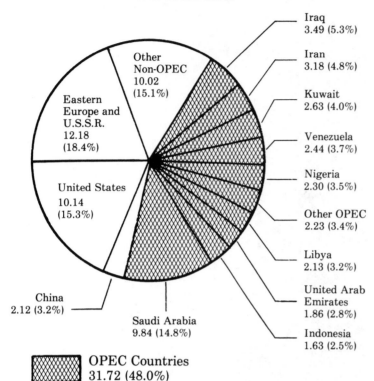

Crude Oil Production
World Total: 66.16

Other Non-OPEC
10.02 (15.1%)

Eastern Europe and U.S.S.R.
12.18 (18.4%)

United States
10.14 (15.3%)

China
2.12 (3.2%)

Saudi Arabia
9.84 (14.8%)

Iraq
3.49 (5.3%)

Iran
3.18 (4.8%)

Kuwait
2.63 (4.0%)

Venezuela
2.44 (3.7%)

Nigeria
2.30 (3.5%)

Other OPEC
2.23 (3.4%)

Libya
2.13 (3.2%)

United Arab Emirates
1.86 (2.8%)

Indonesia
1.63 (2.5%)

OPEC Countries
31.72 (48.0%)

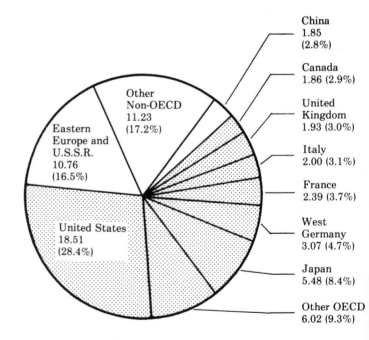

Apparent Consumption of Refined Petroleum Products
World Total: 65.11

Other Non-OECD
11.23 (17.2%)

Eastern Europe and U.S.S.R.
10.76 (16.5%)

United States
18.51 (28.4%)

China
1.85 (2.8%)

Canada
1.86 (2.9%)

United Kingdom
1.93 (3.0%)

Italy
2.00 (3.1%)

France
2.39 (3.7%)

West Germany
3.07 (4.7%)

Japan
5.48 (8.4%)

Other OECD
6.02 (9.3%)

OECD Countries
41.26 (63.4%)

Source: U.S. Department of Energy. Energy Information Administration. <u>1981 Annual Report to Congress</u>, II, 74.

INTERNATIONAL CRUDE OIL FLOW, 1979

(Thousand Barrels Per Day)

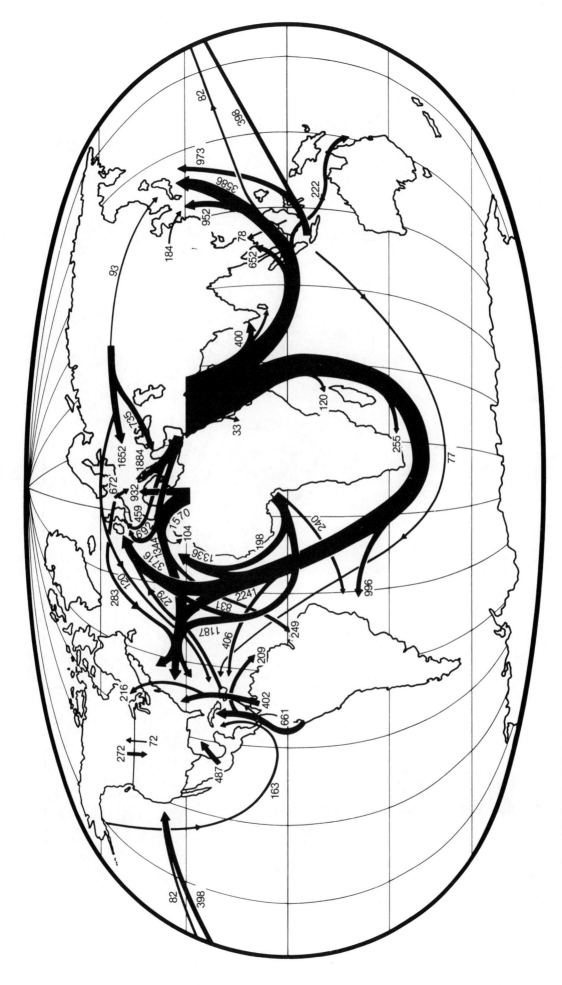

Source: U.S. Department of Energy. Energy Information Administration.
1980 *International Energy Annaul*. September 1981.

Arrows Indicate Origin and Destination
But Not Necessarily Specific Routes

INTERNATIONAL CONSUMPTION OF REFINED PETROLEUM PRODUCTS

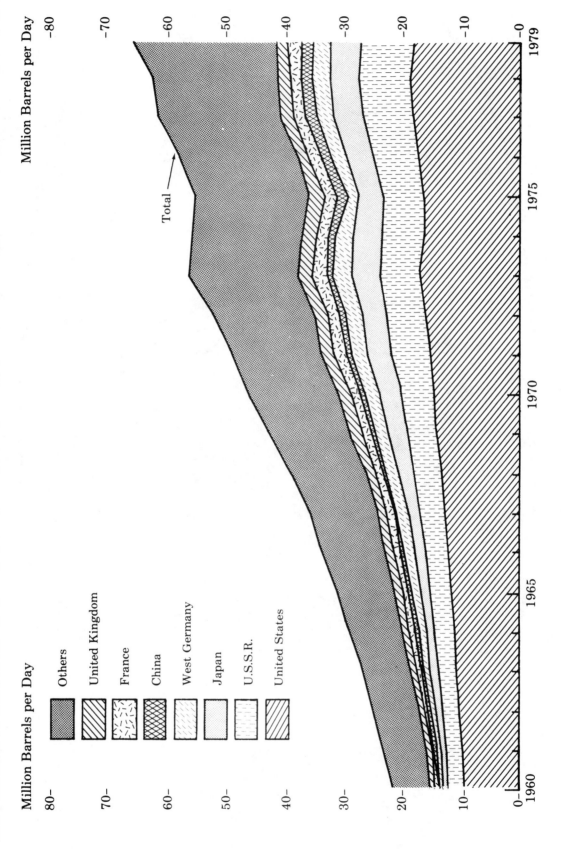

Source: U.S. Department of Energy. Energy Information Administration. 1981 Annual Report to Congress, II, 82.

APPARENT CONSUMPTION OF OIL, FREE WORLD, 1950-1979

(Million Barrels per Day)

Year	United States	Canada	Japan	Europe	Developing Countries	OPEC	Total Free World[b]
1950	6.46	0.32	0.03	1.05	1.98	0.20	10.04
1955	8.46	0.55	0.15	1.96	3.12	0.34	14.58
1960	9.80	0.84	0.64	4.54	2.57	0.67	19.06
1965	11.51	1.14	1.80	8.26	3.63	0.84	27.18
1970	14.70	1.47	4.18	13.58	5.62	1.16	40.71
1971	15.21	1.54	4.41	14.07	6.31	1.29	42.83
1972	16.37	1.69	4.81	14.71	5.80	1.43	44.81
1973	17.31	1.60	5.00	15.15	7.92	1.93	48.91
1974	16.65	1.63	4.87	14.29	9.89	1.99	49.32
1975	16.32	1.60	4.57	12.73	7.53	2.30	45.05
1976	17.46	1.65	4.79	14.03	8.56	2.40	48.89
1977	18.43	1.66	5.02	14.01	8.90	2.64	50.66
1978	18.85	1.70	5.12	13.92	8.17	2.52	50.28
1979	18.50	1.80	5.17	14.50	9.83	2.50	52.30

[a]The 1979 estimates for Europe and OPEC are from U.S. Department of Energy, International Affairs. Other estimates for Europe and OPEC and estimates prior to 1973 for Canada and Japan consist of domestic production of crude oil and natural gas liquids plus net imports of crude oil and petroleum products. Estimates for the Developing Countries were calculated to balance total free world production and apparent consumption by the other regions and include Puerto Rico, Virgin Islands, South Africa, Australia, and New Zealand.

[b]Includes production of crude oil and natural gas liquids and net oil imports from the Centrally Planned Economies.

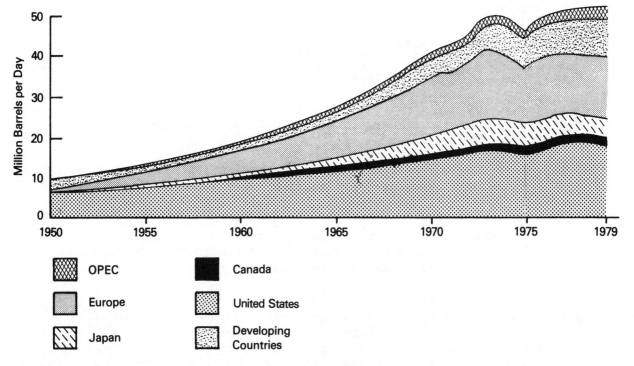

Note:
Data for the years 1951-1954, 1956-1959, 1961-1964, and 1966-1969 are interpolated.

Source: U.S. Department of Energy. 1981.

aspects of "WORLD PETROLEUM AVAILABILITY: 1980-2000"

A Study by the Congressional Office of Technology Assessment (OTA), released October 20, 1980.

Projections of Non-Communist World Oil Supplies*
(in million barrels per day)

	1979 actual	1985	2000
Developed countries			
United States	10.2	7.2- 8.6	4.0- 7.0
Canada	1.8	1.6- 1.8	1.0- 2.0
North Sea countries	2.1	2.8- 4.0	1.7- 3.0
Other developed countries	0.8	0.8	0.8
(A) Total	14.9	13.0-15.5	7.5-13.0
OPEC			
Saudi Arabia	9.8	9.1-11.1	9.1-12.6
Iraq	3.4	2.7- 4.5	3.0- 5.0
Iran	3.0	3.0- 4.0	3.0- 4.0
United Arab Emirates	1.9	1.9- 2.5	1.9- 3.0
Kuwait.	2.6	1.9- 2.4	1.9- 2.4
Other OPEC countries	10.7	9.5-10.5	8.0-10.0
(B) Total	31.4	28.5-35.0	27.0-37.0
Other			
Mexico	1.6	3.0- 4.0	3.5- 5.5
Non-OPEC LDCs	3.5	4.5- 5.0	4.0- 6.0
(C) Total	5.1	7.5- 9.0	7.5-11.5
Net communist trade (negative number indicates imports)			
(D) Total	1.0	(-1.9)-0	(-2)-0
TOTAL (A) + (B) + (C) + (D)	**52.4**	**45-60**	**40-60**

* Includes natural gas liquids.   Note: Totals may not add due to rounding.

1. Non-Communist oil production is expected to drop between now and year 2000.

2. U.S. production is expected to drop. (U.S. production in 1979 was 10.2 millions of (42 gallon) barrels per day. Exxon forecasts it will be 6 MBD in 2000. OTA estimates a range of 4-7 MBD.)

3. The Communist countries may cease being a net exporter of oil to the free world by the early 1980s because of declines in Soviet production.

4. The entry of the Eastern European countries (now more than 80 percent dependent on the Soviet Union for their imported oil) and conceivably the Soviet Union itself as buyers on the world oil market will intensify the pressure on world oil prices and have potentially serious implications for U.S. foreign policy.

5. While it may be physically possible to increase world oil production by perhaps as much as 33 percent by the 1990s, no substantial increases are likely because the countries that must contribute to an increase of this size (such as the Arab OPEC countries or Mexico) have little financial or political incentive to do so and because any attempt to increase production would run into a number of practical as well as political problems.

6. Major additions to the world's known oil supplies will likely come from additional recovery in known fields rather than from new field discoveries. Moreover, experts generally agree that the world distribution of ultimately recoverable oil will not differ significantly from the known distribution today.

NATURAL GAS FLOW DIAGRAM, THE U.S., 1981

(Trillion Cubic Feet)

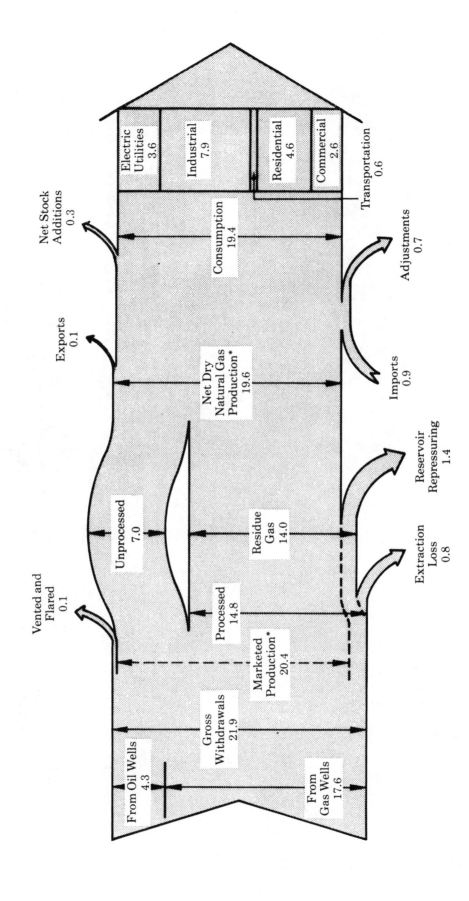

Source: U.S. Department of Energy. Energy Information Administration. *1981 Annual Report to Congress*, II, 101.

*See Glossary.

U.S. CONSUMPTION OF NATURAL GAS BY END-USE SECTOR

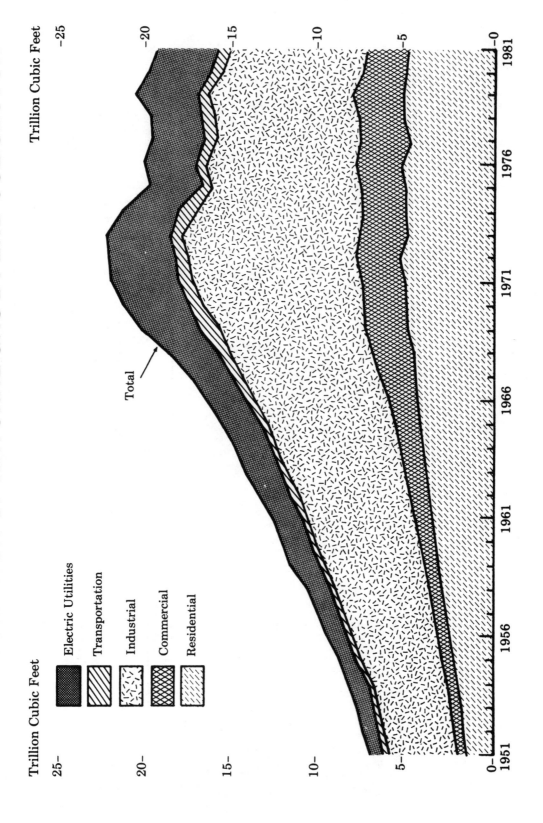

Trillion Cubic Feet

Trillion Cubic Feet

Electric Utilities

Transportation

Industrial

Commercial

Residential

Total

Source: U.S. Department of Energy. Energy Information Administration. 1981 Annual Report to Congress, II, 106.

DISTRIBUTION OF ESTIMATED PROVED RESERVES OF OIL AND GAS AS OF JANUARY 1, 1982

The following is the relevant portion of the footnote for the Table these data were taken from: All reserves figures except those for the U.S.S.R. (and gas for Canada) are reported as proved reserves recoverable with present technology and prices. U.S.S.R. figures are "explored reserves" which includes proved, probable, and some possible. Canadian gas figure, under criteria adopted by Canadian Petroleum Association for the first time in 1980, includes proved and some probable. U.S. reserves numbers based for the first time on Energy Information Administration estimates which ran some 3 billion bbl. and 11 trillion cu.ft. higher than API/AGA estimates for 1979, the last year the latter two associations provided estimates. "Communist" includes Albania, Bulgaria, China, Cuba, Czechoslovakia, East Germany, Hungary, Mongolia, North Korea, Poland, Romania, U.S.S.R., and Vietnam, and Yugoslavia.

Oil (1,000 bbl.)

WORLD	669,709,150
Non-Communist	584,864,150
Communist	84,845,000

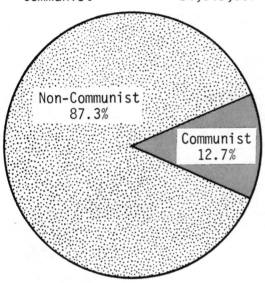

(Gas 10^9 cu. ft.)

WORLD	2,911,346
Non-Communist	1,716,646
Communist	1,194,700

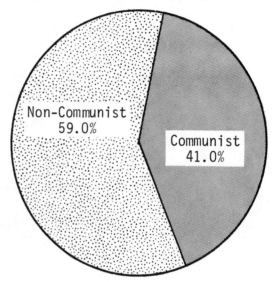

Western Hemisphere	122,067,270
Africa	56,171,630
Middle East	362,839,950
Western Europe	24,634,500
Asia-Pacific	19,150,800

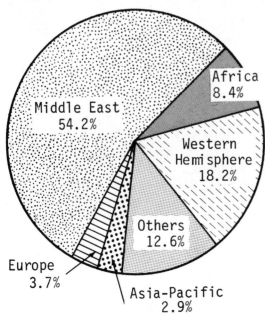

Western Hemisphere	464,223
Africa	211,667
Middle East	762,490
Western Europe	150,650
Asia-Pacific	127,616

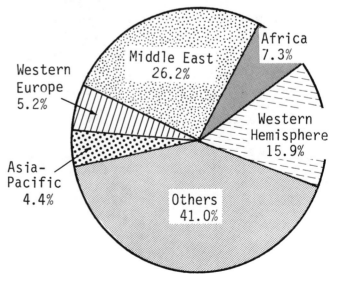

ENERGY GRAPHICS charts based on data from "OGJ Report". Oil and Gas Journal. December 28, 1981.

ESTIMATED PROVED RESERVES OF NATURAL GAS AS OF JANUARY 1, 1982

All reserves figures except those for the U.S.S.R. and Canada are proved reserves recoverable with present technology and prices. U.S.S.R. figures are "explored reserves" which include proved, probable, and some possible. Canadian gas figure, under criteria adopted by Canadian Petroleum Association for the first time in 1980, includes proved and some probable. U.S. reserves numbers based for the first time on Energy Information Administration estimates which ran some 11 cu.ft. higher than AGA estimates for 1979, the last year that association provided estimates.

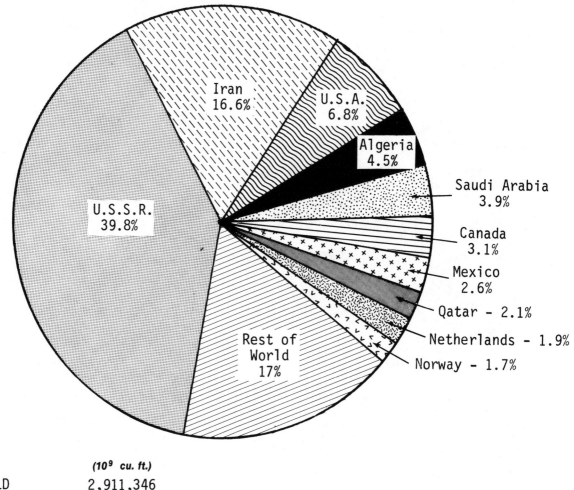

(10⁹ cu. ft.)

	WORLD	2,911,346					
1.	U.S.S.R.	1,160,000	39.8%	12.	Nigeria	40,500	1.4%
2.	Iran	484,000	16.6%	13.	Kuwait	30,500	1.0%
3.	U.S.A.	198,000	6.8%	14.	Indonesia	27,400	0.9%
4.	Algeria	130,900	4.5%	15.	Iraq	27,300	0.9%
5.	Saudi Arabia	114,000	3.9%	16.	United Kingdom	26,000	0.9%
6.	Canada	89,900	3.1%	17.	China	24,400	0.8%
7.	Mexico	75,350	2.6%	18.	Argentina	23,400	0.8%
8.	Qatar	60,000	2.1%	19.	Libya	23,200	0.8%
9.	Netherlands	55,700	1.9%	20.	Abu Dhabi	19,500	0.7%
10.	Norway	49,370	1.7%	21.	Malaysia	19,000	0.7%
11.	Venezuela	47,000	1.6%	22.	Australia	18,700	0.6%

ENERGY GRAPHICS chart based on data in "OGJ Report". <u>Oil and Gas Journal</u>. December 28, 1981.

WORLD NATURAL GAS PRODUCTION (DRY), 1980

(Trillion Cubic Feet)

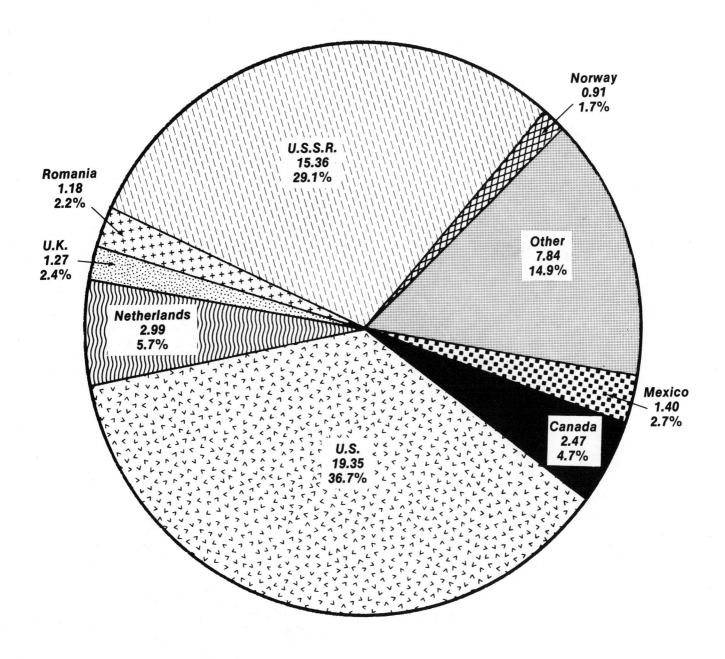

Norway
0.91
1.7%

U.S.S.R.
15.36
29.1%

Romania
1.18
2.2%

Other
7.84
14.9%

U.K.
1.27
2.4%

Netherlands
2.99
5.7%

Mexico
1.40
2.7%

U.S.
19.35
36.7%

Canada
2.47
4.7%

Total 52.77

Sum of components does not equal
total due to independent rounding.

ENERGY GRAPHICS chart based on data from U.S. Department of Energy. Energy Information Administration. 1980 International Annual. September 1981. P.18.

INTERNATIONAL SUPPLY AND DISPOSITION OF NATURAL GAS, 1979

(Trillion Cubic Feet)

Marketed Production
World Total: 52.36

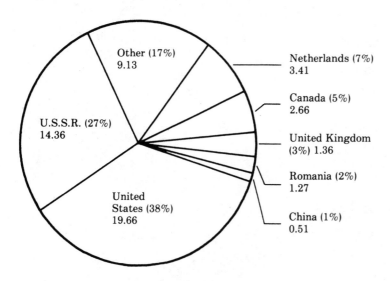

Apparent Consumption
World Total: 51.74

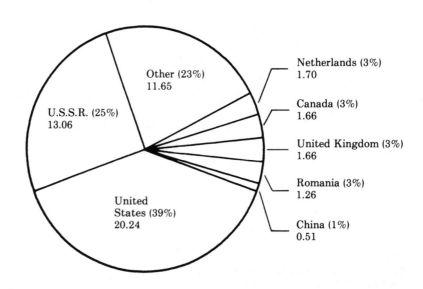

Source: U.S. Department of Energy. Energy
Information Administration. 1981
Annual Report to Congress, II, 110.

MAJOR GAS PIPELINES IN THE U.S.S.R.

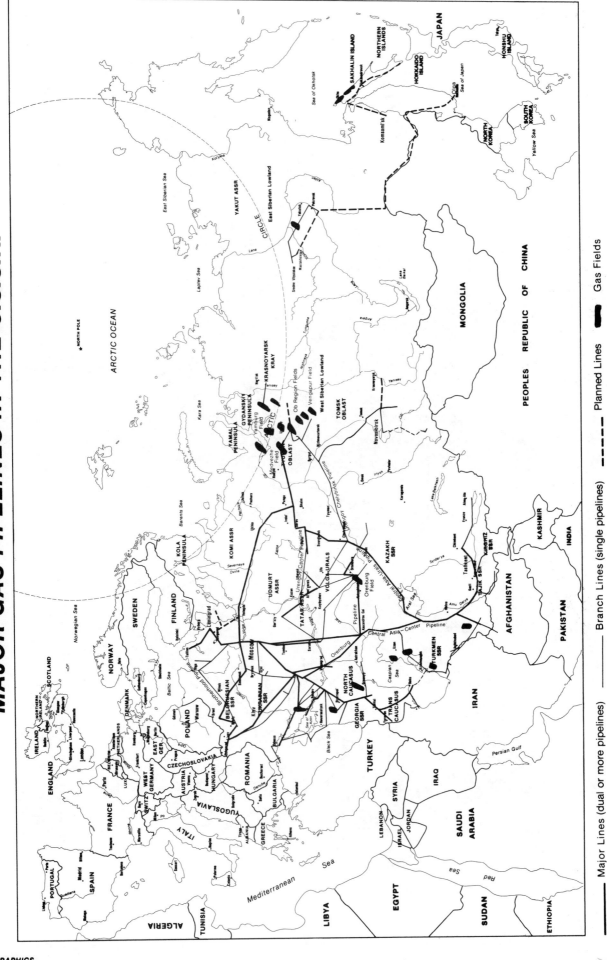

Source: Congress of the United States. Office of Technology Assessment.
Technology & Soviet Energy Availability. November 1981. P.61.

Major Lines (dual or more pipelines) ——— Branch Lines (single pipelines)

– – – Planned Lines ▬ Gas Fields

INTERNATIONAL NATURAL GAS FLOW, 1979

(Billion Cubic Feet)

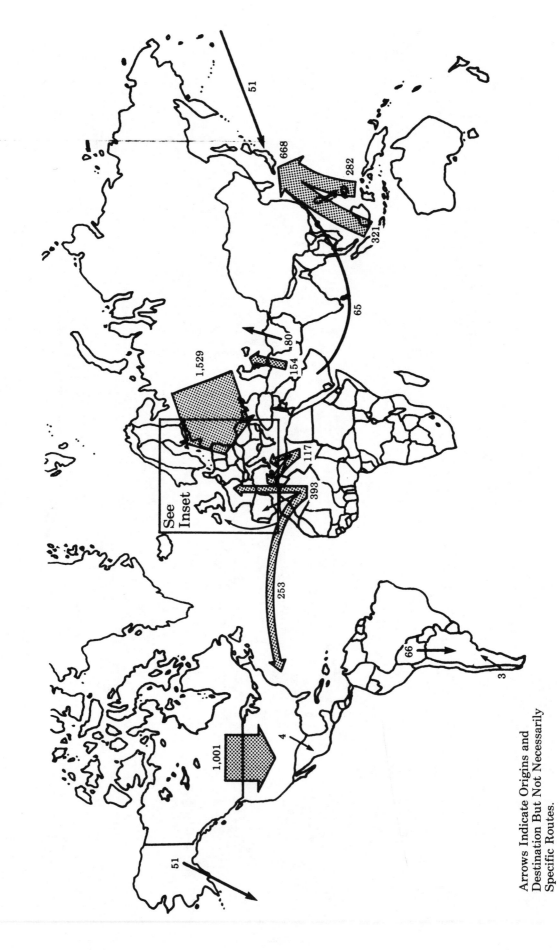

Arrows Indicate Origins and
Destination But Not Necessarily
Specific Routes.

Source: U.S. Department of Energy. Energy Information Administration. 1981 Annual Report to Congress, II, 112.

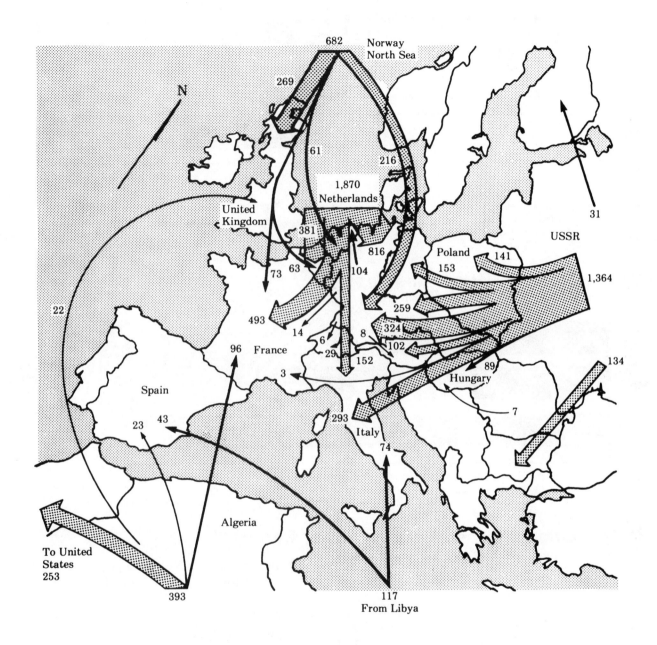

Source: U.S. Department of Energy. Energy Information Administration.
1981 Annual Report to Congress, II, inset to p.112.

WESTERN EUROPE AND SOVIET GAS

Western Europe: Energy Consumption by Type of Fuel
(million barrels per day oil equivalent)

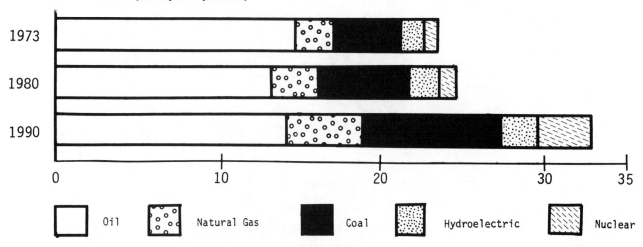

| | Oil | | Natural Gas | | Coal | | Hydroelectric | | Nuclear |

Western Europe: Natural Gas Supply
(million barrels per day oil equivalent)

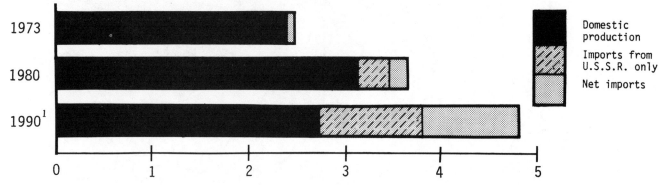

Domestic production

Imports from U.S.S.R. only

Net imports

1 Includes imports from the U.S.S.R. to be received under proposed revision of data.

Western Europe: Dependence on Soviet Gas
(% of total gas and of total energy consumption)

	1979		Original 4bcf/d Project 1990 [1]		Revised 3bcf/d Project 1990	
	Gas	Energy	Gas	Energy	Gas	Energy
West Germany	14	2	29	6	24	5
France [2]	0	0	23-28	4	17-20	3
Italy	29	5	29	5	23	4
Netherlands	0	0	10	4	6	2
Belgium	0	0	32	8	19	5
Austria	59	12	82	18	62	14

1 Based on individual government estimates of total gas and energy consumption.
2 French-contracted volumes of Soviet natural gas were delivered to Italy in exchange for Italian-contracted gas from the Netherlands until February 1980.

Source: U.S. Department of State. Current Policy No. 331. November 1981.

IV. COAL

COALFIELDS OF THE UNITED STATES

(Adapted from U.S. Geological Survey maps; coal types not distinguished in Alaska.)

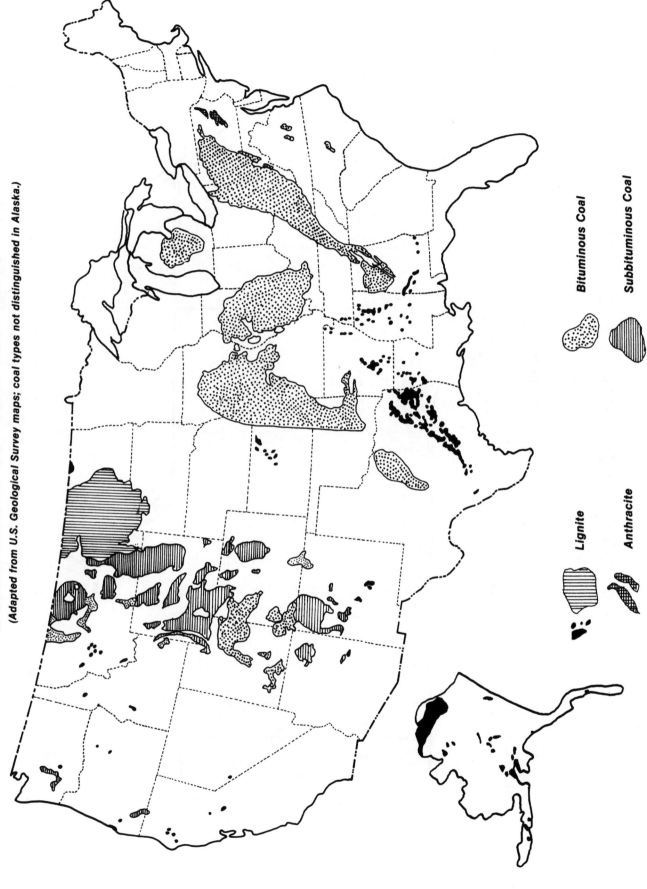

Bituminous Coal

Subbituminous Coal

Lignite

Anthracite

U.S. COAL FLOW DIAGRAM, 1981

(Million Short Tons)

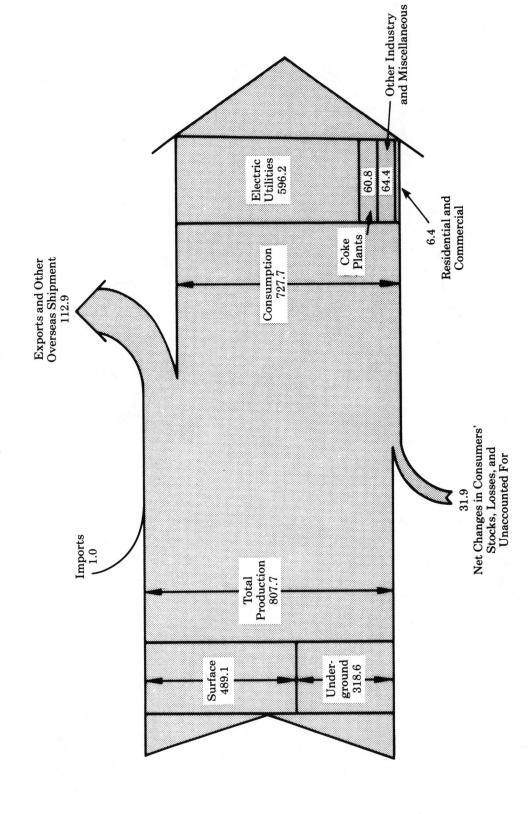

Source: U.S. Department of Energy. Energy Information Administration. 1981 Annual Report to Congress, II, 121.

U.S. COAL PRODUCTION
(Million Short Tons)

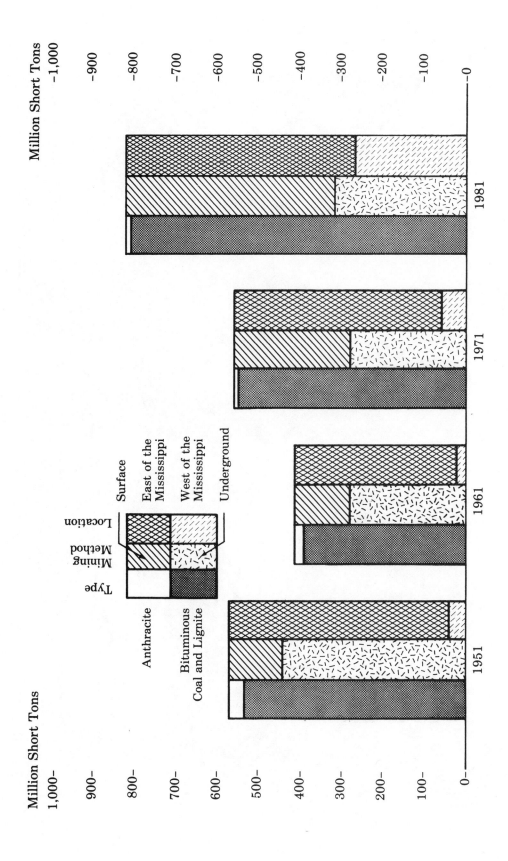

Source: U.S. Department of Energy. Energy Information Administration. 1981 Annual Report to Congress, II, 124.

U.S. COAL CONSUMPTION BY END-USE SECTOR

(Million Short Tons)

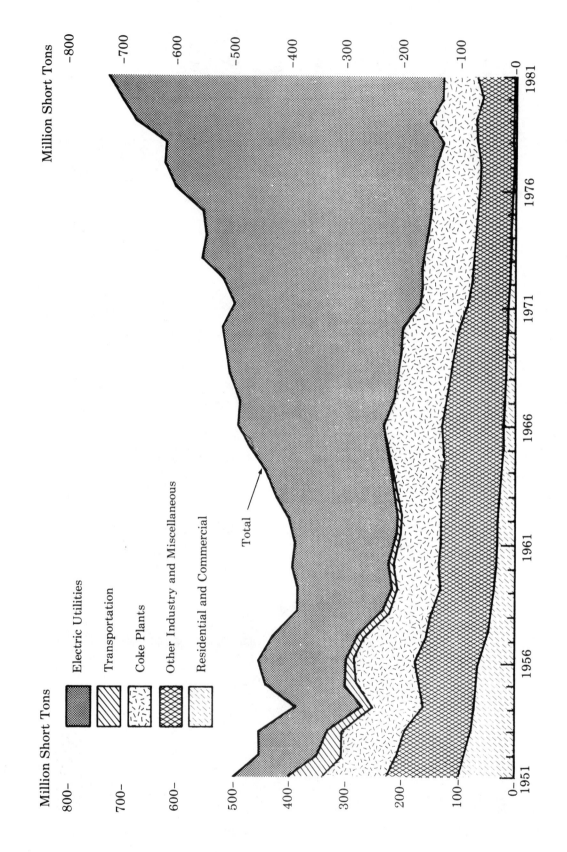

Source: U.S. Department of Energy. Energy Information Administration.
1981 Annual Report to Congress, II, 126.

DEMONSTRATED RESERVE BASE OF COAL BY RANK, REGION, AND POTENTIAL METHOD OF MINING, JANUARY 1, 1979

U.S. Total
474.6 Billion Short Tons

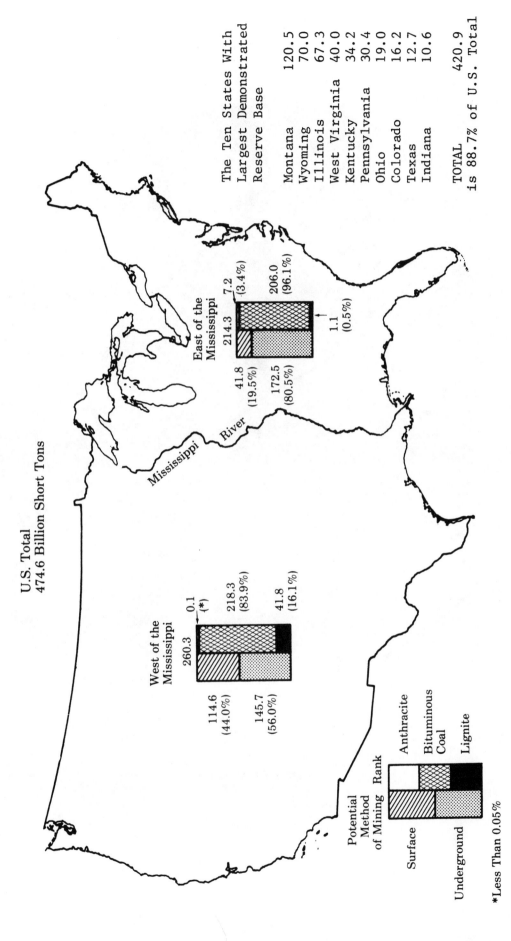

East of the Mississippi

214.3   7.2   206.0
         (3.4%) (96.1%)
                1.1
                (0.5%)

41.8   172.5
(19.5%) (80.5%)

Mississippi River

West of the Mississippi

260.3   0.1   218.3
         (*)   (83.9%)
                41.8
                (16.1%)

114.6   145.7
(44.0%) (56.0%)

Potential Method of Mining   Rank

Anthracite
Bituminous Coal
Lignite

Surface
Underground

*Less Than 0.05%

The Ten States With
Largest Demonstrated
Reserve Base

Montana	120.5
Wyoming	70.0
Illinois	67.3
West Virginia	40.0
Kentucky	34.2
Pennsylvania	30.4
Ohio	19.0
Colorado	16.2
Texas	12.7
Indiana	10.6
TOTAL	420.9

is 88.7% of U.S. Total

Source: U.S. Department of Energy. Energy Information Administration. 1981 Annual Report to Congress, II, 136-137.

WORLD COAL PRODUCTION, 1980

(Million Short Tons)

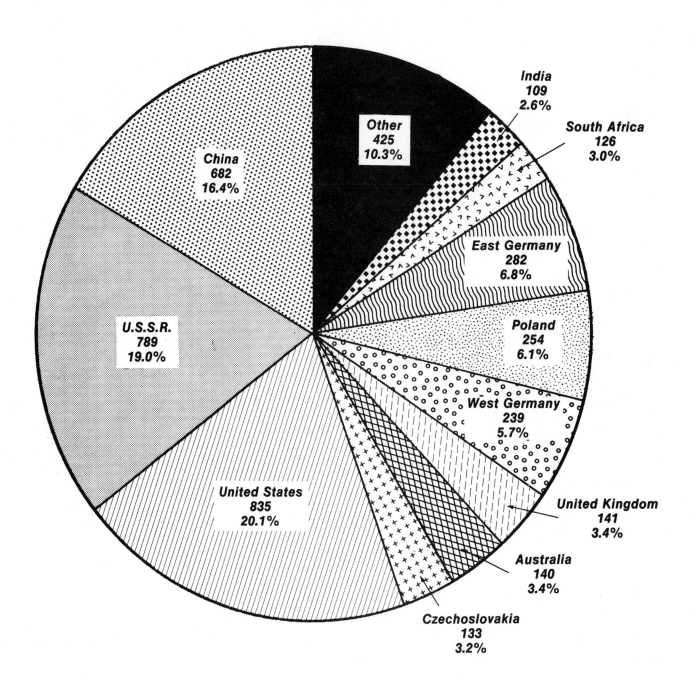

India
109
2.6%

South Africa
126
3.0%

Other
425
10.3%

China
682
16.4%

East Germany
282
6.8%

Poland
254
6.1%

U.S.S.R.
789
19.0%

West Germany
239
5.7%

United Kingdom
141
3.4%

Australia
140
3.4%

United States
835
20.1%

Czechoslovakia
133
3.2%

World Total: 4,155

ENERGY GRAPHICS chart based on data from U.S. Department
of Energy. Energy Information Administration. <u>1980 International Energy Annual</u>. September 1981. P.20.

ESTIMATED INTERNATIONAL RECOVERABLE RESERVES OF COAL, 1979

(Billion Short Tons)

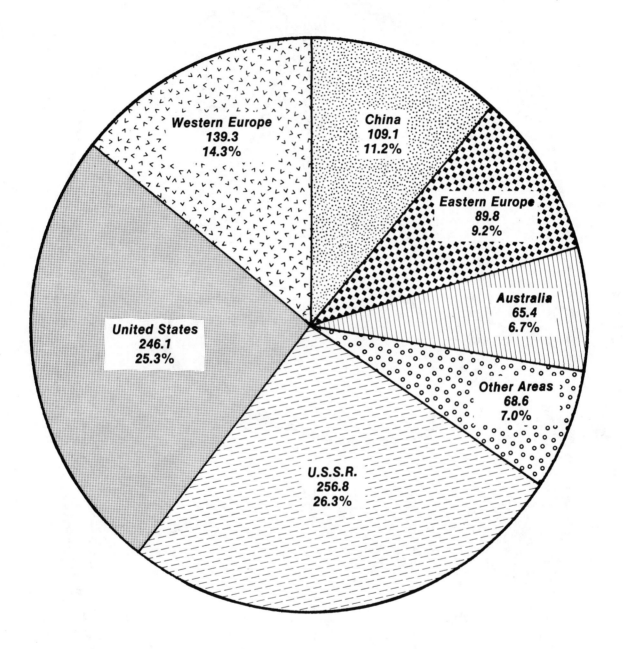

Western Europe
139.3
14.3%

China
109.1
11.2%

Eastern Europe
89.8
9.2%

Australia
65.4
6.7%

United States
246.1
25.3%

Other Areas
68.6
7.0%

U.S.S.R.
256.8
26.3%

World Total: 975.1

ENERGY GRAPHICS chart based on data from U.S. Depart of Energy. Energy Information Administration. 1981 Annual Report to Congress, II, 140.

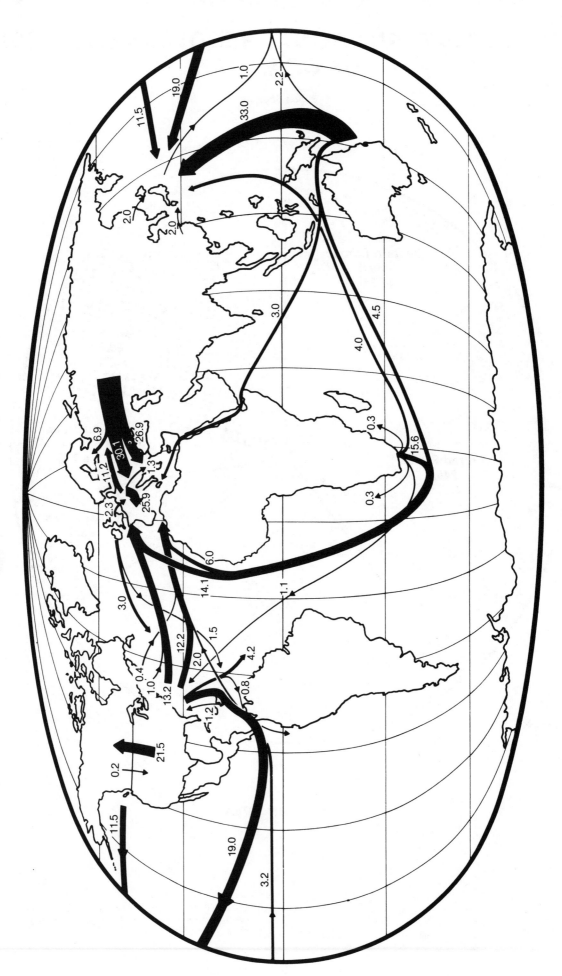

INTERNATIONAL COAL FLOW, 1979

(Million Short Tons of Coal Equivalent)

Arrows Indicate Origin and Destination
But Not Necessarily Specific Routes

Source: U.S. Department of Energy. Energy Information Administration. 1980 International Energy Annual. September 1981. P.79.

U.S. COAL EXPORTS BY COUNTRY OF DESTINATION
(Million Short Tons)

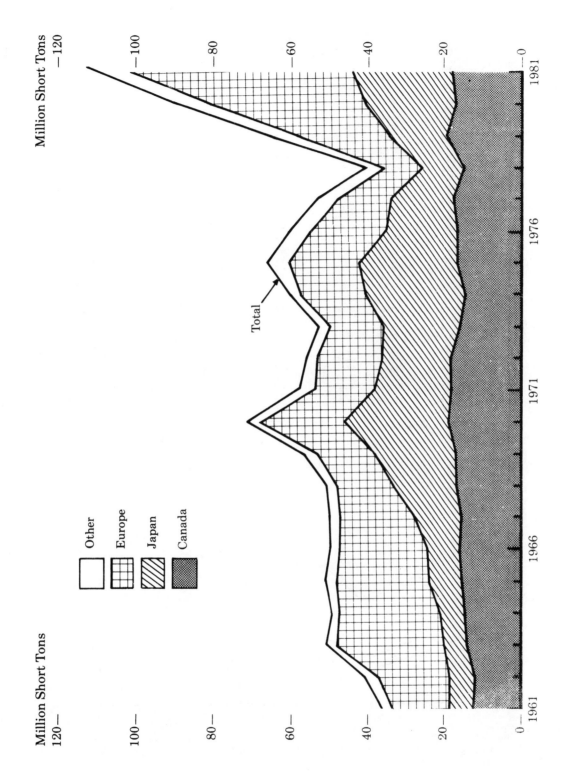

Source: U.S. Department of Energy. Energy Information Administration. 1981 Annual Report to Congress, II, 128.

HISTORY AND PROJECTION OF U.S. COAL EXPORTS

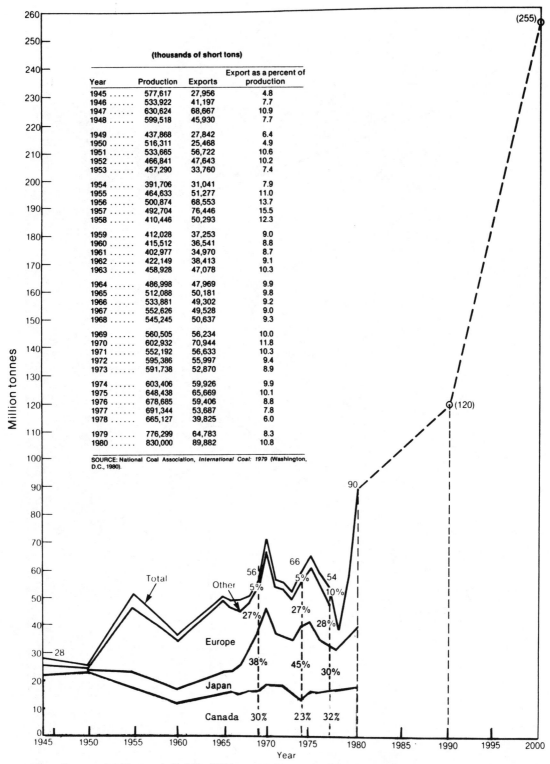

(thousands of short tons)

Year	Production	Exports	Export as a percent of production
1945	577,617	27,956	4.8
1946	533,922	41,197	7.7
1947	630,624	68,667	10.9
1948	599,518	45,930	7.7
1949	437,868	27,842	6.4
1950	516,311	25,468	4.9
1951	533,665	56,722	10.6
1952	466,841	47,643	10.2
1953	457,290	33,760	7.4
1954	391,706	31,041	7.9
1955	464,633	51,277	11.0
1956	500,874	68,553	13.7
1957	492,704	76,446	15.5
1958	410,446	50,293	12.3
1959	412,028	37,253	9.0
1960	415,512	36,541	8.8
1961	402,977	34,970	8.7
1962	422,149	38,413	9.1
1963	458,928	47,078	10.3
1964	486,998	47,969	9.9
1965	512,088	50,181	9.8
1966	533,881	49,302	9.2
1967	552,626	49,528	9.0
1968	545,245	50,637	9.3
1969	560,505	56,234	10.0
1970	602,932	70,944	11.8
1971	552,192	56,633	10.3
1972	595,386	55,997	9.4
1973	591,738	52,870	8.9
1974	603,406	59,926	9.9
1975	648,438	65,669	10.1
1976	678,685	59,406	8.8
1977	691,344	53,687	7.8
1978	665,127	39,825	6.0
1979	776,299	64,783	8.3
1980	830,000	89,882	10.8

SOURCE: National Coal Association, *International Coal: 1979* (Washington, D.C., 1980).

Note: Steam coal at 30 percent of total in 1980 is expected to grow to 78 percent by 2000.

SOURCES: History-Coal data book. Projection-ICE Task Force with constant 1980 metallurgical coal added.

From table and chart, pp.6 & 34, <u>Coal Exports and Port Development</u>. April 1981. By Congress of the United States. Office of Technology Assessment.

COAL-PRODUCING COUNTIES IN THE UNITED STATES

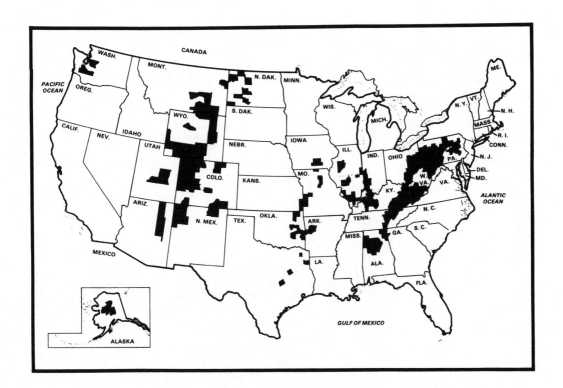

In the 1970's, U.S. coal production fluctuated from 561 million tons to 776 million tons. From 1890 to 1978, the U.S. produced 44.6 billion tons. Of the 438.3 billion tons of Demonstrated Coal Reserve Base (January 1, 1976) in the U.S., it is estimated that at least 219 billion tons are recoverable. (Beyond the Demonstrated Coal Base, are an additional 1.3 trillion tons of identified resources and 2.2 trillion tons of estimated undiscovered resources.

Source: U.S. Department of Energy. 1980.

International Coal Production, 1980
(million short tons)   (preliminary)

WORLD TOTAL: 4,155

U.S.A.	835	20.1%
U.S.S.R.	789	19.0
PRC	682	16.4
Germany, GDR	282	6.8
Poland	254	6.1
Germany, FRG	239	5.7
U.K.	141	3.4
Australia	140	3.4
Czech.	133	3.2
So. Africa	126	3.0
India	109	2.6
Other	425	10.3

Source: U.S. Department of Energy. Energy Information Administration. 1980 International Energy Annual. September 1981. P.20.

Estimated International Recoverable Reserves of Coal, 1979
(billion short tons)

WORLD TOTAL: 975.1

U.S.S.R.	256.8	26.3%
U.S.A.	246.1	25.3
Western Europe	139.2	14.3
PRC	109.1	11.2
Eastern Europe	89.8	9.2
Australia	65.4	6.7
Other	68.6	7.0

Source: U.S. Department of Energy. Energy Information Administration. 1981 Annual Report to Congress, II, 140.

SOVIET COAL BASINS

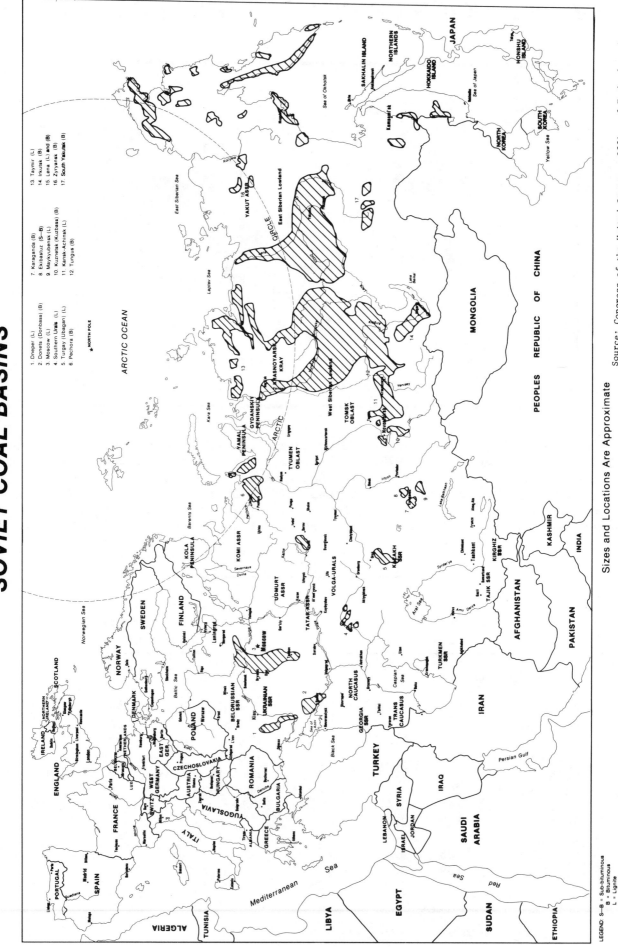

1 Dneper (L)
2 Donets (Donbass) (B)
3 Moscow (L)
4 Southern Urals (L)
5 Turgay (Ubagan) (L)
6 Pechora (L)

7 Karaganda (B)
8 Ekibastuz (S–B)
9 Maykyubensk (L)
10 Kuznetsk (Kuzbass) (B)
11 Kansk-Achinsk (L)
12 Tungus (B)

13 Taymir (L)
14 Irkutsk (B)
15 Lena (L) and (B)
16 Zyryansk (B)
17 South Yakutsk (B)

Sizes and Locations Are Approximate

Source: Congress of the United States. Office of Technology Assessment. Technology & Soviet Energy Availability. November 1981. P.85.

LEGEND S–B = Sub-bituminous   B = Bituminous   L = Lignite

NUMBER OF FEDERAL COAL ACRES UNDER LEASE BY BUSINESS ACTIVITY CATEGORY, 1950-1980

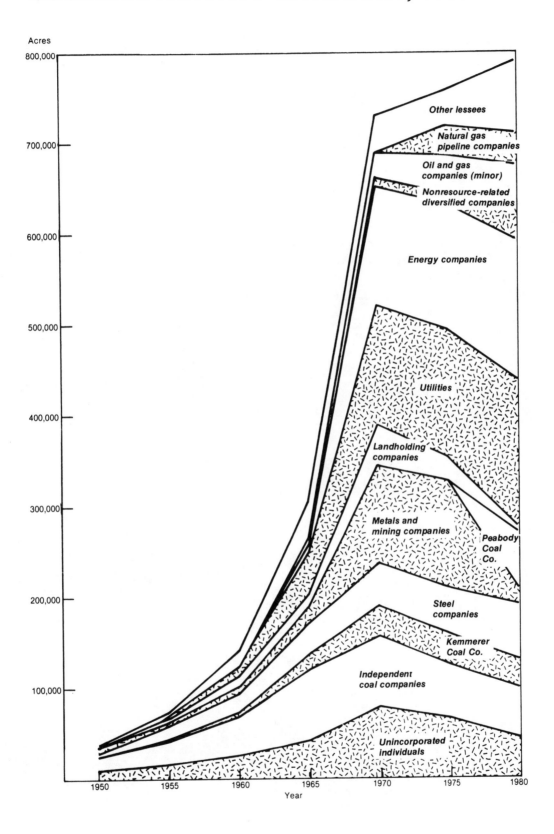

Source: United States Congress. Office of
Technology Assessment. 1981.

V. NUCLEAR

WORLD COMMERCIAL NUCLEAR POWER GENERATION, 1970 AND 1980

(Excludes Communist countries)

	Number of Reactors		Electricity Generated (bil. kWh)		Gross Capacity (1,000 kW)	
	1970	1980	1970	1980	1970	1980
WORLD.	64	208	73.9	617.8	15,186	128,847
U.S.	15	74	23.2	265.2	5,211	56,529
Argentina.	--	1	--	2.3	--	357
Belgium.	1	3	0.3	12.5	11	1,744
Canada .	1	9	0.9	40.4	220	5,588
Germany, Fed. Rep.	4	11	5.3	43.7	907	8,996
Great Britain.	27	33	26.5	37.2	4,783	9,012
Finland.	--	4	--	7.0	--	2,296
France .	4	22	5.7	61.2	1,606	15,412
India.	2	4	2.2	2.9	400	860
Italy.	3	4	3.3	2.2	631	1,490
Japan.	3	22	3.3	81.0	828	15,117
Netherlands.	1	2	0.4	4.2	55	529
Pakistan .	--	1	--	.1	--	137
South Korea.	--	1	--	3.5	--	587
Spain.	1	3	0.9	5.2	160	1,117
Sweden .	1	8	n.a.	26.7	12	5,770
Switzerland.	1	4	1.9	14.3	364	2,034
Taiwan .	--	2	--	8.2	--	1,272

Source: U.S. Department of Commerce. 1981 Statistical Abstract of the United States. P.595. Table 1027.

STATUS OF U.S. NUCLEAR POWERPLANTS, DECEMBER 31, 1981

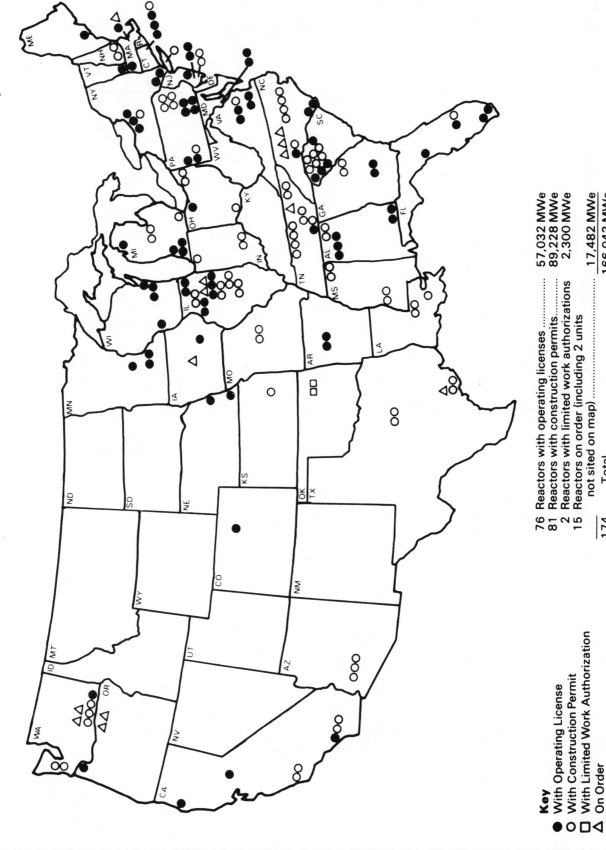

Source: Atomic Industrial Forum, Inc.

76	Reactors with operating licenses	57,032 MWe
81	Reactors with construction permits	89,228 MWe
2	Reactors with limited work authorizations	2,300 MWe
15	Reactors on order (including 2 units not sited on map)	17,482 MWe
174	Total	166,042 MWe

June 30, 1981

Key
● With Operating License
○ With Construction Permit
□ With Limited Work Authorization
△ On Order

NUCLEAR POWER REACTORS IN OPERATION AND UNDER CONSTRUCTION AT THE END OF 1981

Country	In operation		Under construction	
	Number of units	Total MW(e)	Number of units	Total MW(e)
Argentina	1	335	2	1 292
Belgium	3	1 664	4	3 807
Brazil			3	3 116
Bulgaria	3	1 224	2	1 408
Canada	11	5 494	14	9 751
"China, Republic of"	3	2 159	3	2 765
Cuba			1	408
Czechoslovakia	2	800	6	2 520
Finland	4	2 160		
France	30	21 595	26	28 585
German Democratic Republic	5	1 694	4	1 644
Germany, Federal Republic of	14	8 606	10	10 636
Hungary			2	816
India	4	809	4	880
Italy	4	1 417	3	1 999
Japan	24	14 994	12	9 973
Korea, Republic of	1	564	8	6 869
Mexico			2	1 308
Netherlands	2	501		
Pakistan	1	125		
Philippines			1	620
Romania			1	660
South Africa			2	1 842
Spain	4	1 973	11	10 142
Sweden	9	6 415	3	3 025
Switzerland	4	1 940	1	942
Union of Soviet Socialist Republics	35	14 036	25	24 260
United Kingdom	32	7 627	9	5 533
United States of America	75	57 008	79	87 217
Yugoslavia	1	632		
World Total	272	153 772	238	222 018

Construction in Austria and Iran has been interrupted and plants in these countries are not included.

Source: International Atomic Energy Agency
Bulletin, March 1982. P.2.

NUCLEAR

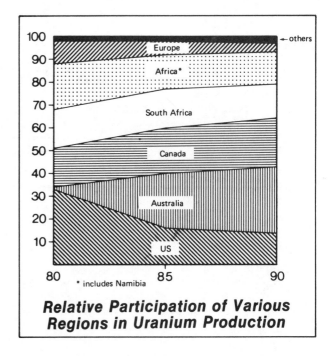

Relative Participation of Various Regions in Uranium Production

* includes Namibia

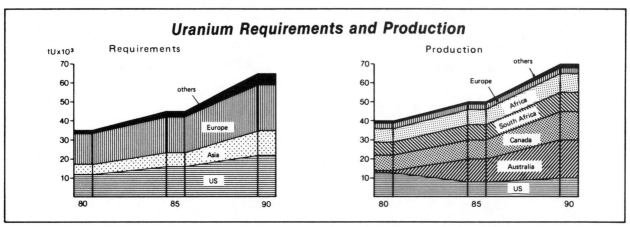

Uranium Requirements and Production

Estimates of Total and Nuclear Electrical Generating Capacity by Main Country Groups (GWe)

Country group	1980 Total	Nuclear	%	1985 Total (average)	Nuclear	%	1990 Total (average)	Nuclear	%
OECD North America	710	57	8	890	130	15	1065	150	14
OECD Europe	440	45	10	580	105	18	735	150	20
OECD Pacific	180	15	8	255	25	10	340	50	15
Centrally Planned Europe	370	16	4	545	35	6	745	75	10
Total for industrialized countries	1700	133	8	2270	295	13	2885	425	15
Asia	130	3	2	235	10	4	400	20	5
Latin America	100	0.3	0.3	130	3	2	180	10	6
Africa and Middle East	65	—	—	80	2	3	120	3	3
Total for developing countries	295	3	1	445	15	3	700	33	5
World total	1995	136	7	2715	310	11	3585	458	13

Source: International Atomic Energy Agency
Bulletin, September 1981.

NUCLEAR ELECTRICITY GENERATION BY NON-COMMUNIST COUNTRIES

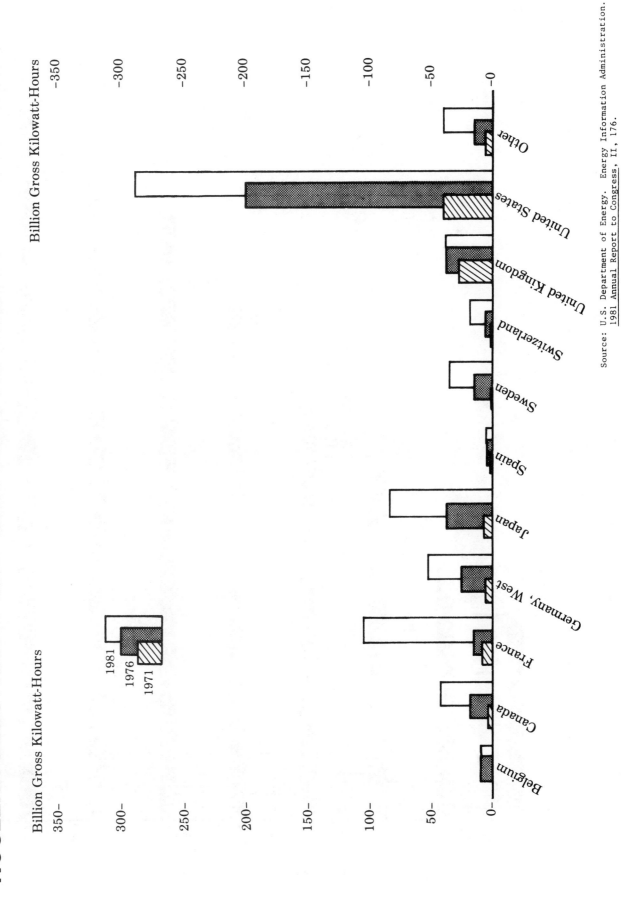

Billion Gross Kilowatt-Hours

Source: U.S. Department of Energy. Energy Information Administration. 1981 Annual Report to Congress, II, 176.

ENERGY GRAPHICS
64 Washburn Ave., Wellesley, MA 02181

WORLD NUCLEAR ELECTRIC POWER PRODUCTION (NET), 1980

(Billion Kilowatt Hours)

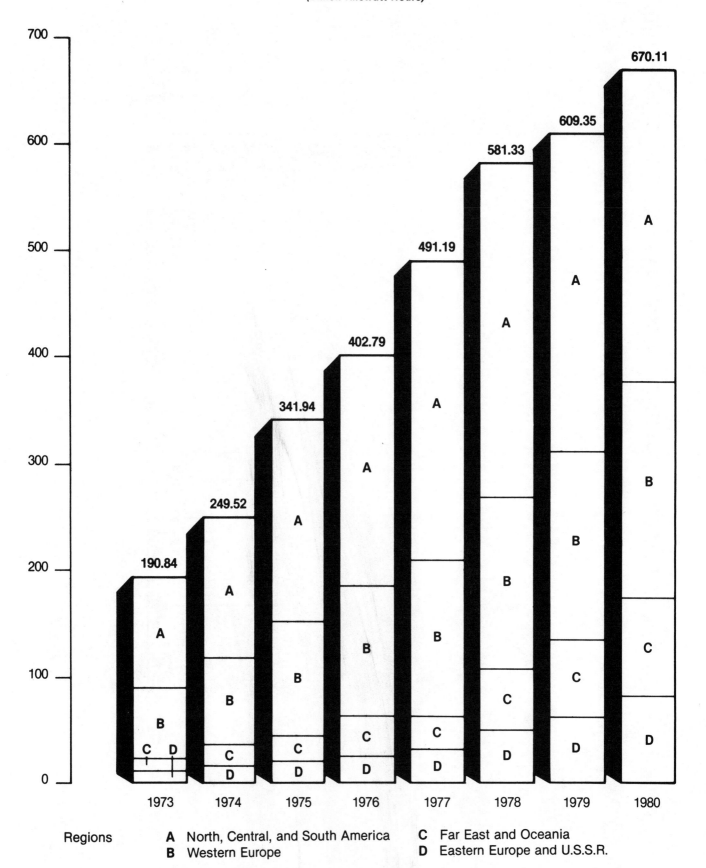

Regions   **A** North, Central, and South America   **C** Far East and Oceania
    B Western Europe   **D** Eastern Europe and U.S.S.R.

Source: U.S. Department of Energy. Energy Information Administration.
1980 International Energy Annual. September 1981. P.25.

ENERGY GRAPHICS
64 Washburn Ave., Wellesley, MA 02181

INSTALLED NUCLEAR ELECTRIC GENERATING CAPACITY[1]

(Thousand Megawatts)

Year	Total	United States	Japan	Western Europe	Canada	Communist Countries	Other Countries
1960	0.9	0.4	0	0.5	0	0	0
1970	15.7	6.0	0.8	7.3	0.2	1.0	0.4
1974	55.0	29.4	3.7	13.6	2.5	4.8	1.0
1975	70.5	37.4	5.0	17.8	2.5	6.8	1.0
1976	81.3	40.8	7.1	21.6	2.5	8.3	1.0
1977	94.6	47.6	7.6	25.1	4.0	9.3	1.0
1978	106.7	50.1	11.0	27.4	4.8	11.2	2.2
1979	122.7	51.3	14.5	35.3	5.5	12.9	2.8
1980	131.7	52.2	14.5	40.8	5.5	15.9	2.8

1 Data are as of the end of the year and presented on a net basis except for
Communist countries which are gross capacities.

Source: CIA. Handbook of Economic Statistics. 1981.

U.S. NUCLEAR POWERPLANT CAPACITY AND ELECTRICITY PRODUCTION

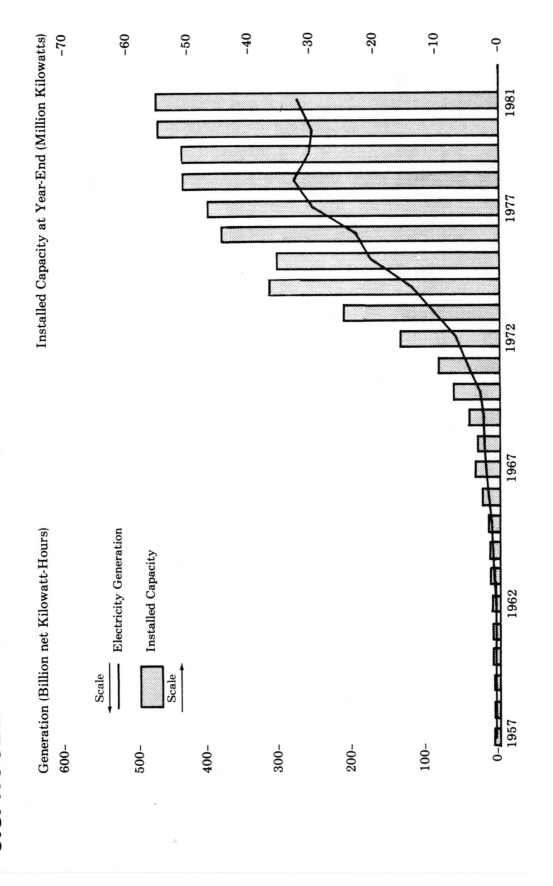

Installed Capacity at Year-End (Million Kilowatts)

Generation (Billion net Kilowatt-Hours)

Scale — Electricity Generation

Scale ▢ Installed Capacity

Source: U.S. Department of Energy. Energy Information Administration. 1981 Annual Report to Congress., II, 170.

U.S. URANIUM RESOURCES, JANUARY 1, 1981

**Forward Cost
$30 per Pound or Less**

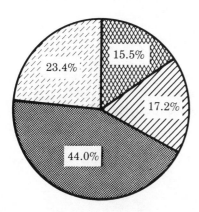

2,012 Thousand
Short Tons of U$_3$O$_8$

**Forward Cost
$50 per Pound or Less**

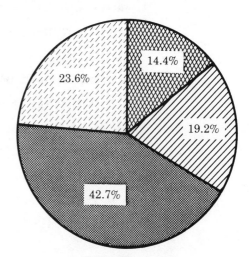

3,336 Thousand
Short Tons of U$_3$O$_8$

 Reserves   Speculative Resources

Probable Resources   Possible Resources

Note: Quantities scaled in proportion to area.

Source: U.S. Department of Energy. Energy Information Administration.
1981 Annual Report to Congress, II, 172.

U.S.S.R. 1985 GOAL FOR ELECTRIC GENERATING CAPACITY

1,550 billion to 1,600 billion kWh from all sources

Of which 220 billion to 225 billion kWh to be from nuclear generation stations

This amount would be approximately 14%[1] of total electricity generated

To achieve these goals, 24,000 to 25,000 MW of additional nuclear generating capacity would have to be installed during Five Year Plan of 1981-1985

Total installed nuclear capacity by the end of 1985 would then be about 38,000 MW [2]

(The U.S. had 52,300 MW of nuclear capacity as of January 1980)

---

1 "Draft of the Main Directions of Economic and Social Development of the U.S.S.R. for 1981-1985 and the Period of 1990," Izvestiya, Dec. 2, 1980, pp. 2-7.
2 This assumes that the Soviet Union had 13,460 MW of effective nuclear power generating capacity, as of Dec. 31, 1980. Addition of 24,000 to 25,000 MW of capacity during the Eleventh FYP would result in total nuclear capacity at the end of 1985 of 37,460 to 38,460 MW. OTA has used the approximate midpoint of this range -- 38,000 MW.

Source: Congress of the United States. Office of Technology Assessment. Technology & Soviet Energy Availability. November 1981. P.120.

THE NUCLEAR FUEL CYCLE

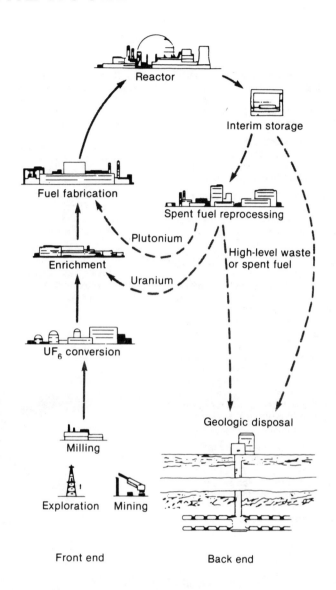

Reactor

Interim storage

Fuel fabrication

Spent fuel reprocessing

Plutonium

High-level waste or spent fuel

Enrichment

Uranium

UF$_6$ conversion

Milling

Geologic disposal

Exploration   Mining

Front end   Back end

Fuel cycle today
Prospective "closed" fuel cycle

The commercial nuclear fuel cycle includes activities for pre-
paring and using reactor fuel and for managing spent fuel and
other radioactive wastes produced in the process. It was
originally intended that spent fuel be stored for 6 months in
water-filled basins at reactor sites to dissipate thermal heat
and allow decay of short-lived fission products. The spent fuel
would then be reprocessed and the resultant liquid high-level
waste solidified and disposed of in a Federal repository. Since
no repository has been developed and no commercial reprocessing is being done, spent fuel will remain in storage until
decisions about how to close the nuclear fuel cycle are made.

Source: Council on Environmental Quality.

FUSION ENERGY . . .

. . . what is it?

Nuclear energy can be made available in two ways. One way is through *fission,* the process that is the basis for today's nuclear power plants. Fission involves splitting the atomic nuclei of "heavy" elements such as uranium or plutonium into two parts, a process that releases energy.

The second way is through *fusion,* which occurs when the nuclei of light elements, such as hydrogen, are fused together to form a heavier element. A fusion reaction produces much more energy than is needed to make the reaction occur.

Fusion is the process by which the sun and stars generate energy. In a sense, our sun is a gigantic fusion reactor that continuously converts hydrogen to helium through a complex chain of fusion reactions, and releases the resulting energy in the form of sunlight.

. . . why is it important?

As the world's energy needs have increased, the limits of the earth's resources have become apparent. Fossil fuels such as gas and oil can no longer be depended upon as long-term abundant sources of energy. Fusion energy can be both safe and environmentally acceptable, and the supply of its basic fuel, *deuterium* (a form of hydrogen), is virtually unlimited: it can be obtained from ordinary water. Each gallon of seawater contains enough deuterium to produce energy equivalent to about 300 gallons of gasoline. (The other component of the fuel mixture, *tritium*—another form of hydrogen—is constantly generated within the fusion device when neutron particles resulting from fusion reactions strike a surrounding blanket containing the element *lithium.*)

Fusion is being pursued not as an imminent alternative to current energy sources, but as a possible long-term solution to at least part of the world's energy problem. The successful production of useful power from fusion—which researchers believe will come in the early decades of the twenty-first century—can mean the generation of enough electricity for the entire world for tens of thousands of years.

Some of the problems

Producing controlled fusion on earth is an extremely complex problem that has not yet been completely solved. There are three basic requirements that must be fulfilled in order for fusion reactions to occur rapidly enough that significant energy can be produced:

- *Temperature.* Atomic nuclei are positively charged, and normally repel one another. This electrical repulsion can be overcome only by heating the nuclei to the point that they travel at the rate of more than a million miles an hour— requiring temperatures of about 100 million degrees.

- *Density.* There must be a high enough concentration of particles in the fusion fuel that many reactions can occur in a short time.

- *Time.* For a significant number of fusion reactions to take place, the fuel must hold its energy—in technical jargon, it must be *confined*—for about one full second.

In brief, researchers must confine a highly unstable gas containing a large number of reacting particles at very high temperatures for a specific length of time.

Because it is so hot, the fusion fuel (which is a *plasma*—a superhot, ionized gas) cannot be contained by any known material. To solve this problem, much of this country's current fusion research centers on the *magnetic confinement* approach to fusion, which involves holding the plasma within powerful magnetic fields inside the fusion device. (The other major approach is *inertial confinement,* based on the very rapid heating of a fuel pellet with laser or particle beams.) Given the restless nature of a plasma, this task has been described by some researchers as being like trying to hold jello with rubber bands! However, fusion researchers have made considerable progress in the past five years toward meeting all three criteria, and believe they can be achieved in the near future.

A current major concern of fusion researchers is whether a fusion power system can be economical. Ultimately, the test of a fusion reactor will be whether it can produce continuous, reliable power at a cost that is reasonable compared to other energy sources.

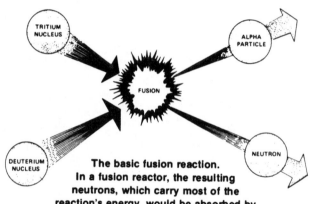

The basic fusion reaction.
In a fusion reactor, the resulting neutrons, which carry most of the reaction's energy, would be absorbed by a surrounding "blanket" that would slow them down and release their energy as heat to a working fluid, generating steam to drive an electric turbine-generator. Also, interactions of the neutrons with lithium in the blanket would generate more tritium fuel.

Source: Oak Ridge National Laboratory Operated by Union Carbide Corporation's Nuclear Division for the U.S. Department of Energy.

VI. NEW AND RENEWABLE

KNOWN AND POTENTIAL HYDROTHERMAL RESOURCES

The KGRA refers to known geothermal resources area.

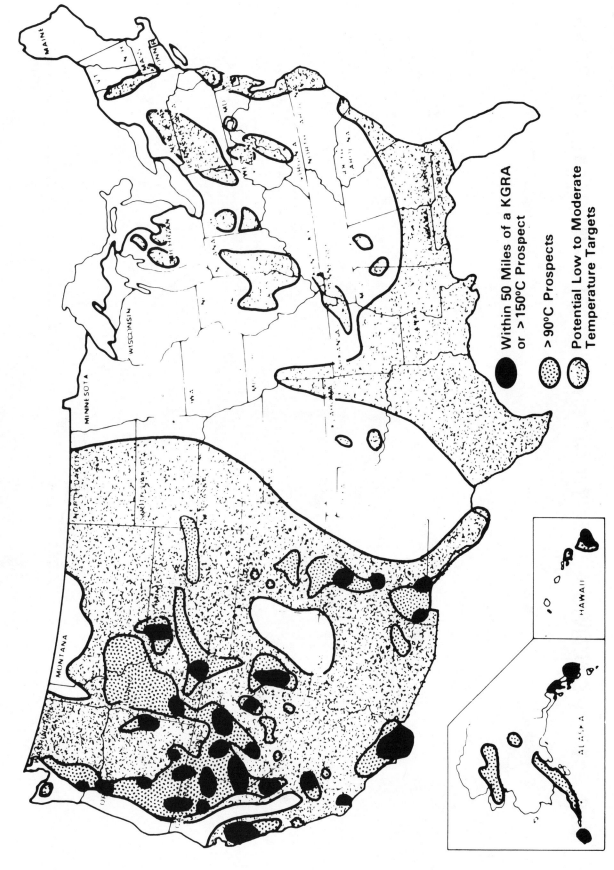

Within 50 Miles of a KGRA or >150°C Prospect

> 90°C Prospects

Potential Low to Moderate Temperature Targets

Source: U.S. Department of Energy. 1981.

U.S. PRODUCTION OF ELECTRICITY FROM GEOTHERMAL SOURCES

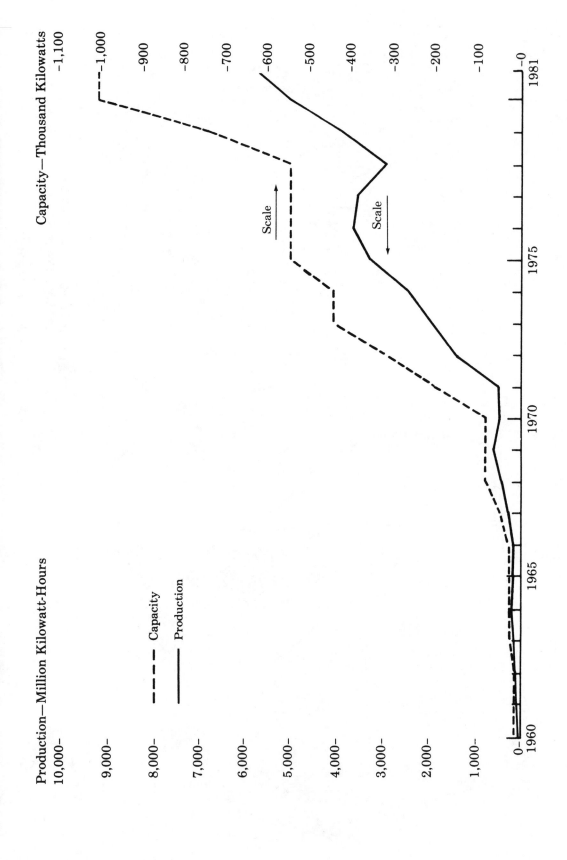

Production—Million Kilowatt-Hours

Capacity—Thousand Kilowatts

- - - Capacity

——— Production

Source: U.S. Department of Energy. Energy Information Administration. 1981 Annual Report to Congress, II, 180.

ILLUSTRATIVE COMPARISON OF TYPICAL
CORN-TO-ETHANOL/COAL-TO-METHONAL PLANTS

Coal-to-Ethanol Plant	Coal-to-Methanol Plant

Annual (Daily) Output

50-Million Gallons per Year (3,600 Barrels per Day)	693-Million Gallons per Year (50,000 Barrels per Day)

Capital Cost *

$90 to 100-Million	$2 to 4 Billion

Required Land Area for Plant

25 to 50 Acres	300 to 600 Acres

Work Force

75 to 100 Persons	350 to 400 Persons

Process Water Requirements

4 to 5 Gallons per Gallon of Output	6 to 7 Gallons per Gallon of Output **

Output Over Life of Plant

1 Billion Gallons (23-Million Barrels per 20-Year Production Cycle)	13.9 Billion Gallons (330-Million Barrels per 20-Year Production Cycle)

Construction Time

2 to 4 Years	6 to 8 Years

Daily Raw Material Requirement

1,649 Tons of Corn (58,900 Bushels)	23,000+ Tons of Bituminous Coal

* Does not include first year operating costs.
** Water use for bituminous or sub-bituminous coal. Water requirements for lignite will be significantly less.

Source: Final Report. U.S. National Alcohol Fuels Commission. Washington 1981. FUEL ALCOHOL: An Energy Alternative for the 1980s. P. 92.

BIOMASS RESOURCE BASE FOR ALCOHOL PRODUCTION, 1990 AND 2000

Raw Material	Potential Alcohol Production (Billions of Gallons) Ethanol	Methanol	Raw Material	Potential Alcohol Production (Billions of Gallons) Ethanol	Methanol
Cellulose			Grain		
Wood			1990	4.0	--
1990	3.2	6.2	2000	4.0	--
2000	1.9	3.7	Sugar Crops		
Municipal Solid Wastes			1990	3.0	--
1990	3.7	9.8	2000	5.0	--
2000	4.3	11.5	Sludge		
Crop Wastes			1990	--	0.1
1990	1.5	3.4	2000	--	0.1
2000	1.5	3.4	Food Wastes		
Subtotal for Cellulose (Wood, MSW, Crop Wastes)			1990	0.5	--
			2000	0.6	--
1990	8.4	19.4	TOTAL		
2000	7.7	18.6	1990	15.9	19.5
			2000	17.3	18.7

* Cellulose may be converted to either ethanol or methanol. The figure given for each alcohol under any cellulose category is a maximum number, assuming that the particular resource is used only to produce that alcohol. A given amount of cellulose may be converted to a larger volume of methanol than ethanol. Therefore, the total volume of alcohol that may be produced from a given cellulose resource will range from the minimum (ethanol) figure to the maximum (methanol) figure.

** Excludes potential contribution from silvicultural energy farms.

POTENTIAL ALCOHOL YIELD OF GRAINS AND POTATOES

	Corn	Grain Sorghum	Wheat (Avg)	Rice	Potatoes
Starch	71.1%	71.1%	67.3%	65.1%	16.6%
Cost ($) per lb. Starch	0.070-0.087	0.067-0.090	0.099-0.125	0.378-0.407	0.361-0.602
Typical Gallon Ethanol/Acre	272	149	82	197	372

ESTIMATED U.S. ETHANOL PRODUCTION CAPACITY
(Cumulative Totals of January 1 of Year Indicated)

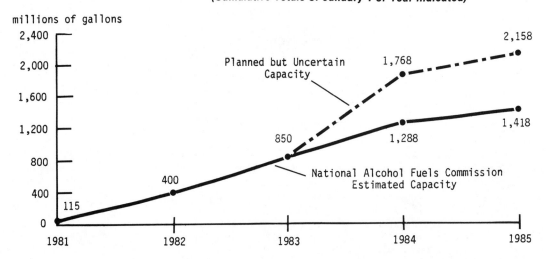

Source: Final Report. U.S. National Alcohol Fuels Commission. Washington 1981. FUEL ALCOHOL: An Energy Alternative for the 1980s. Pp.59,69,70.

WORLD HYDROELECTRIC POWER PRODUCTION (NET),1980

(Preliminary)   **(Billion Kilowatt Hours)**

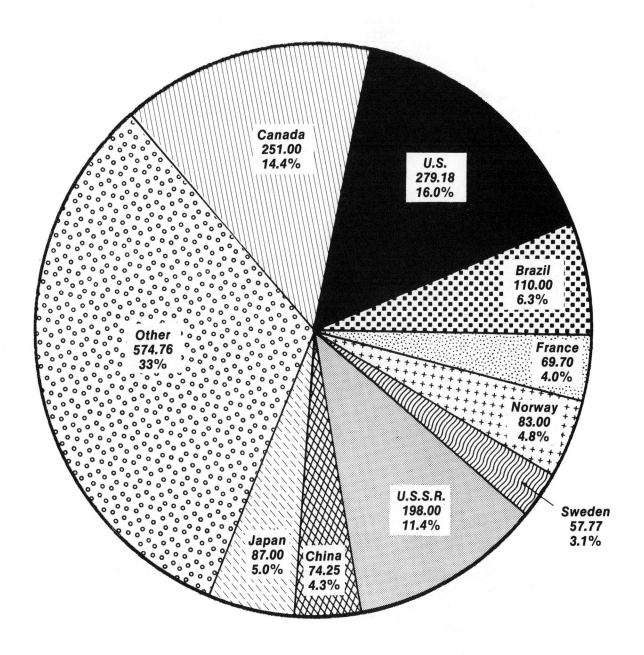

Canada
251.00
14.4%

U.S.
279.18
16.0%

Brazil
110.00
6.3%

France
69.70
4.0%

Norway
83.00
4.8%

Other
574.76
33%

Sweden
57.77
3.1%

U.S.S.R.
198.00
11.4%

Japan
87.00
5.0%

China
74.25
4.3%

World Total: 1,743.66

*Sum of components does not equal
total due to independent rounding.*

ENERGY GRAPHICS chart based on data from U.S. Department
of Energy. Energy Information Administration. 1980
International Energy Annual. September 1981. P.22.

HYDROPOWER

Hydropower Potential and Use, by Region, 1980

Region	Technically Exploitable Potential	Exploited Resources	Share of Potential Exploited
	----- megawatts -----		-percent-
Asia	610,100	53,079	9
South America.	431,900	34,049	8
Africa	358,300	17,184	5
North America.	356,400	128,872	36
U.S.S.R.	250,000	30,250	12
Europe	163,000	96,007	59
Oceania.	45,000	6,795	15
WORLD.	2,200,000	363,000	17

Selected Countries Obtaining Most of Their Electricity from Hydropower, 1980

Country	Share (%)	Country	Share (%)
Ghana	99	Brazil	87
Norway	99	Portugal	77
Zambia	99	New Zealand	75
Mozambique	96	Nepal	74
Zaire	95	Switzerland	74
Sri Lanka	94	Austria	67
		Canada	67

Status of Hydropower Development, by Region, 1980

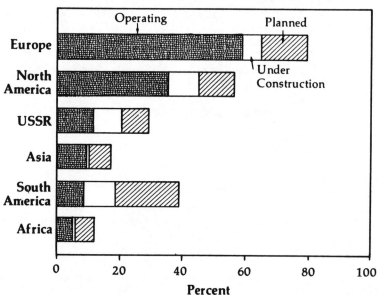

Source: World Energy Conference. Also its Survey of Energy Resources. And UN, World Energy Supplies. As reprinted in Worldwatch Institute Paper No. 44. June 1981.

U.S. PRODUCER SHIPMENTS OF SOLAR COLLECTORS BY TYPE OF COLLECTOR AND APPLICATION, 1980

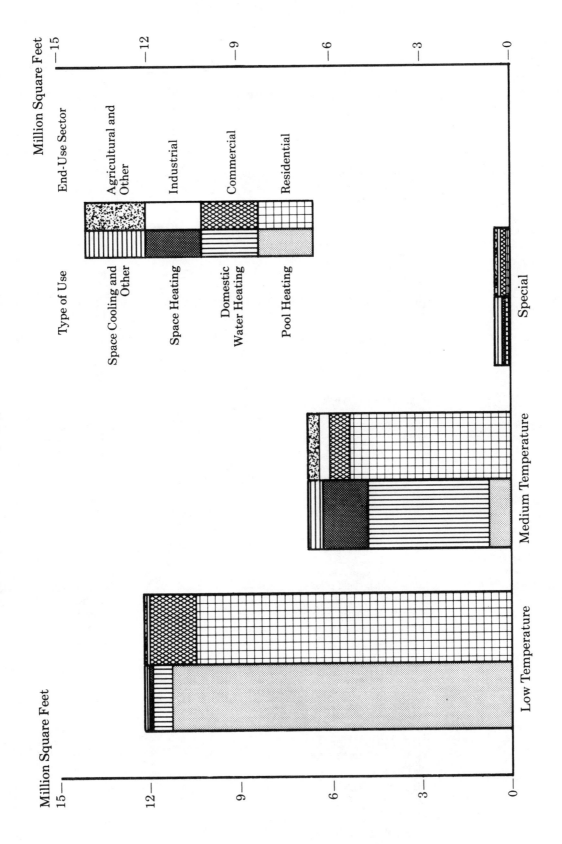

Source: U.S. Department of Energy. Energy Information Administration. 1981 Annual Report to Congress, II, 184.

MAJOR POTENTIAL SOLAR POWER SATELLITES ENVIRONMENTAL IMPACTS

System component[a] characteristics	Environmental impact	Public health and safety	Occupational health and safety
Power transmission Microwave	—[b]Ionospheric heating could disrupt telecommunications —[b]Microwave bioeffects and reflected light (on plants, animals, and airborne biota) —[b]Interference with satellite communications, terrestrial communications, radar, radio, and optical astronomy	—Low-level chronic exposure to microwaves —Psychological effects of microwave beam as weapon. —Adverse esthetic effects on appearance of night sky	—Protective clothing required for terrestrial worker —Accidental exposure to high intensity beam in space
Lasers	—Tropospheric heating —Beam may incinerate birds and vegetation —[b]Interference with optical astronomy, some interference with radio astronomy	—Ocular hazard? —Psychological effects of laser as weapon are possible —Adverse esthetic effects on appearance of night sky are possible	—Ocular and safety hazard
Mirrors	—[b]Tropospheric heating —24-hr light affects growing cycles of plants and circadian rhythms of animals —[b]Interference with optical astronomy	—Ocular hazard? —Psychological effect of 24-hr sunlight might occur —[b]Adverse esthetic effects on appearance of night sky are possible	—Ocular hazard?
Transportation and Space Operation Launch and recovery	—Ground cloud might pollute air and water and cause possible weather modification —[b]Water vapor and other launch effluents could deplete ionosphere and enhance airglow. Disruption of communications and satellite surveillance? —[b]Formation of noctilucent clouds in stratosphere and mesosphere —[b]Emission of water vapor could alter natural hydrogen cycle —[b]Effect of COTV argon ions on magnetosphere and plasmasphere —Noise	—Noise (sonic boom) may exceed EPA guidelines —Ground cloud might affect air quality —Accidents-catostrophic explosion near launch site, vehicle crash, toxic materials	—[b]Space worker's hazards: ionization radiation (potentially severe), weightlessness life support failure, long stay in space, contruction accidents, psychological stress, acceleration —Terrestrial worker's hazards: noise, transportation accidents

continued ⟶

MAJOR POTENTIAL SOLAR POWER SATELLITES ENVIRONMENTAL IMPACTS continued

System component[a] characteristics	Environmental impact	Public health and safety	Occupational health and safety
Terrestrial activities Mining	—Land disturbance —Increase of air and water pollution —Solid waste generation —Strain on production capacity of gallium, arsenide, sapphire, silicon, graphite fiber, tungsten and mercury	—Toxic material exposure —Increase of air and water pollution —Land use disturbance	—Air and water pollution —Toxic materials exposure —Noise
Manufacturing	—Increase of air and water pollution —Solid wastes	—Increase of air and water pollution —Solid wastes —Exposure to toxic materials	—Toxic materials exposure —Noise
Construction	—Land disturbance —Local increase of air and water pollution	—Land disturbance —Local increase of air and water pollution	—Noise —Local increase of air and water pollution —Accidents
Receiving antenna	—[b]Land use and siting	—[b]Land use-reduced property value, esthetics, vulnerability	—Waste heat
High-voltage transmission lines (not unique to SPS)	—[b]Land use and siting —Ecosystem: bioeffects of powerlines uncertain	—Exposure to high intensity EM fields	—Exposure to high intensity EM fields

[a]Impacts based on SPS systems as currently defined and do not account for offshore receivers or possible mitigating system modifications.

[b]Research priority.

Source: United States Congress. Office of Technology Assessment. Solar Power Satelites. (Summary.) 1981.

WORLD USE OF
NEW AND RENEWABLE SOURCES

	Now	**Year 2000**
	in billion (10⁹) kWh	
Solar	2-3	2,000-5,000
Geothermal	55	1,000-5,000
Wind	2	1,000-5,000
Tidal	0.4	30-60
Wave	0	10
Thermal gradient of the sea	0	1,000
Biomass	550-700	2,000-5,000
Fuelwood	10,000-12,000	15,000-20,000
Charcoal	1,000	2,000-5,000
Peat	20	1,000
Draught animals	30 (in India)	1,000
Oil shale	15	500
Tar sands	130	1,000
Hydropower	1,500	3,000

Conventional sources include fossil fuels (oil, natural gas, coal) plus nuclear power.

New and renewable sources include solar, geothermal, wind, tidal, wave, thermal gradient of the sea, biomass, fuelwood, charcoal, peat, draught animals, oil shale, tar sands, hydropower.

Source: United Nations.

UNITED STATES ANNUAL AVERAGE WIND POWER

Pacific Northwest Laboratories
Operated for the U.S. Department of Energy
by Battelle Memorial Institute

MAP DESCRIPTION

This wind resource map, produced by Pacific Northwest Laboratory for the U.S. Department of Energy, is based on a synthesis of 12 regional assessments covering the United States and its territories. For each region, wind resource assessments are presented in the form of an atlas. The atlases depict in graphic, tabular, and narrative form, the wind resource on a regional and state level.

Local terrain features may interact with the winds and cause the wind power to vary as much as 50% to 100% from the assessment values. The terrain may be local areas of higher wind power in regions estimated to have low wind power; conversely, some local areas may have lower wind power than that shown in this assessment depicting the degree of certainty of these assessment values should always be used in combination with this wind resource map.

The analyses of mean wind power apply to terrain features that are well exposed to the wind, such as plains, tablelands, hilltops, ridge crests and mountain summits. In wooded or urban areas, the assessment values represent large clearings and other locations free of obstructions to the wind. For coastal areas, the offshore hand represents estimates for exposed coastal locations.

The physical characteristics of the land-surface form affect the number of wind turbines that can be located and spaced. For example, over most of the land area in a flat plain may be favorably exposed to the wind, whereas in mountainous terrain only the ridge crests may be favorably exposed. In exposed places, the estimated mean wind power generally exceeds 1000 ft. mountainous areas and prominent ridge crests are confined in heavy black lines with black marks. Within these areas, wind resource estimates are for exposed ridge crests and mountain summits.

The classes of wind power density used in the assessment are defined in the table to the left. Vertical extrapolation of wind speed to 10 m and 50 m is based on the 1/7 power law, generally applicable to well-exposed sites. The decrease of air density with altitude requires a higher mean wind speed to achieve a given wind power. To obtain the same wind power classes a given wind speed is estimated to be about 1% higher than shown in the table for every 1000 ft of elevation.

LOCATIONS NOT SHOWN ON THIS MAP

Areas not shown on this map but included in the regional assessments are the U.S. Virgin Islands and Pacific Islands. Wind power resource in the U.S. Virgin Islands have a wind power class 4 and higher have areas with class 5, 6 and 7. Pacific Islands include: Midway, Wake, and Johnston Islands, the Northern Marianas and American Samoa. Only the more windward wind areas of Guam and American Samoa. Portions of the U.S. Virgin Islands indicate high wind resource.

CLASSES OF WIND POWER DENSITY

WIND POWER CLASS	10m (33 ft) WIND POWER W/m²	10m SPEED m/s	10m SPEED mph	50m (164 ft) WIND POWER W/m²	50m SPEED m/s	50m SPEED mph
1	0	0	0	0	0	0
2	100	4.4	9.8	200	5.6	12.5
3	150	5.1	11.5	300	6.4	14.3
4	200	5.6	12.5	400	7.0	15.7
5	250	6.0	13.4	500	7.5	16.8
6	300	6.4	14.3	600	8.0	17.9
7	400	7.0	15.7	800	8.8	19.7
	1000	9.4	21.1	2000	11.9	26.6

RIDGE CREST ESTIMATES (LOCAL RELIEF > 1000 FT)
FOR MORE INFORMATION, SEE MAP DESCRIPTION

PUERTO RICO

0   20   40 KILOMETERS
0   20   40 MILES

PRINCIPAL HAWAIIAN ISLANDS

0   20   40 MILES
0   20   40 KILOMETERS

ALASKA

0   100   200 MILES
0   100   200 KILOMETERS

0   100   200   300 MILES
0   100   200   300 KILOMETERS

Courtesy of BATTELLE. Pacific Northwest Laboratories. Richland, Washington. Reprinted with permission.

TYPICAL ANNUAL ENERGY OUTPUT FOR SMALL WIND SYSTEMS

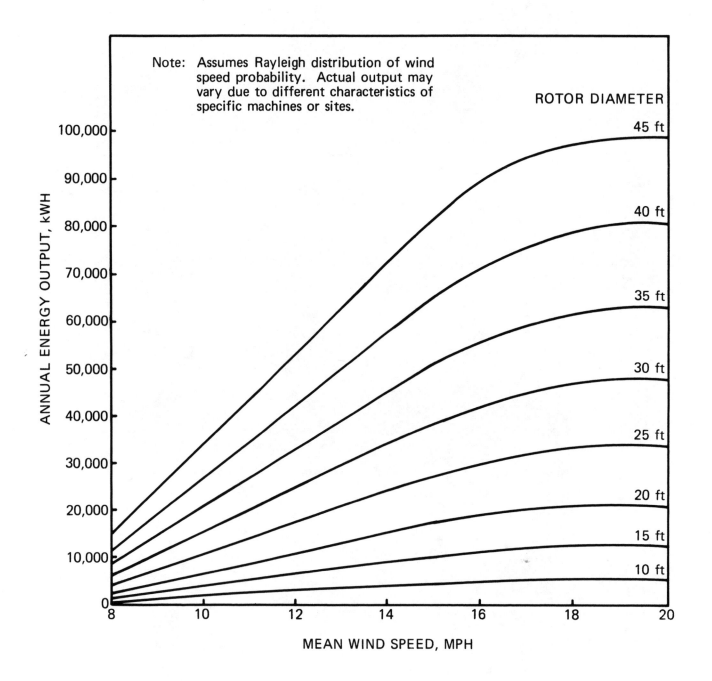

Note: Assumes Rayleigh distribution of wind speed probability. Actual output may vary due to different characteristics of specific machines or sites.

ROTOR DIAMETER

45 ft
40 ft
35 ft
30 ft
25 ft
20 ft
15 ft
10 ft

ANNUAL ENERGY OUTPUT, kWH

100,000
90,000
80,000
70,000
60,000
50,000
40,000
30,000
20,000
10,000
0

MEAN WIND SPEED, MPH

8   10   12   14   16   18   20

Source: U.S. Department of Energy. Energy Information Administration. 1980 Annual Report to Congress, II, 7.

WORLD'S FIRST MEGAWATT-SIZE WIND FARM

Three giant wind turbines at the Goodnoe Hills near the Columbia River Gorge (upper left) in Washington State are now working together for the first time in the nation's only multi-megawatt wind turbine cluster. To date they have produced more than two million kilowatt hours of electricity.

The 2500 kilowatt Mod-2 wind turbines were designed and built by Boeing Engineering and Construction Co. under a progam sponsored by the Department of Energy and managed by NASA's Lewis Research Center. The output of the three machines, enough to provide electricity to 2000 average homes, is being fed into the Northwest power grid operated by the Bonneville Power Administration.

The 350-foot-tall machines, about one third mile apart, are designed to begin producing power in winds of 14 miles per hour. They reach rated output in winds of 28 mph, and when the wind exceeds 45 mph, the blade tips automatically feather, shutting down the system.

Unit No. 1 (foreground) was damaged during tests in 1981, but was back on line in April 1982. The wind farm is now in full cluster operation, with testing and research underway to determine how well the machines work with the transmission system, how they work together, and how they affect the environment.

Courtesy of Boeing Engineering and Construction Company,
a Division of The Boeing Company. Reprinted with permission.

NEW AND RENEWABLE TECHNOLOGY SUMMARY-IN-BRIEF

	Solar Collectors: Low-Temperature Passive	Solar Collectors: Low-Temperature Active	Solar Collectors: Intermediate- and High-Temperature	Photovoltaics	Biomass Combustion	Biomass Conversion
Present State of Development Scale: R&D Stage / Technically Ready / Demonstration Stage / Commercially Ready	**Commercially Ready** Techniques used for centuries; some lack of experience for new techniques; need for dissemination of design approaches and performance data.	**Demonstration Stage** Domestic water systems are commercially available. Space heating systems are being demonstrated but need cost reduction. Space cooling requires R&D and demonstration.	**Demonstration Stage** Central receiver systems and parabolic dish systems need demonstration. Parabolic trough systems are available commercially.	**Commercially Ready** For electrical needs in remote areas. Technically ready for residential electric applications with grid backup. R&D stage for utility use.	**Commercially Ready** Oldest known technology. Wood from nearby forests and aggregated, dry organic wastes are the most commercial. Experience in commercial operations is needed.	**Demonstration Stage** Anaerobic digesters, chemical reduction, gasification, and liquefaction systems are developed. Smaller units are commercial; larger units need demonstration.
Resource/Raw Materials Availability Scale: Fair / Good / Very Good / Excellent	**Excellent** All materials are conventional and readily available.	**Excellent** All materials are readily available.	**Good** No constraint on fabrication materials. Land availability may be a problem for industrial process heat.	**Excellent** Refining capacity for specific feedstocks will have to be increased but it is not a constraint.	**Excellent** Availability is regional but not a constraint.	**Excellent** Biomass can be produced on marginal lands; availability of water may be a limiting factor.
Fossil Fuel Displacement Potential (Year 2000) Scale: Fair / Good / Very Good / Excellent	**Good** Heating of residential and commercial buildings requires a great deal of low-temperature heat. Passive systems can readily supply 50 to 100 percent of this required heat and save fossil fuels.	**Good** Potential displacement of electricity and oil for domestic water heating, space heating in buildings, and industrial process heat totals about 5 exajoules.	**Very Good** Systems are broadly applicable for process heat, electricity, and other industrial applications where fossil fuels provide much of the present power.	**Fair** Attractive in dispersed applications but displacement is not large.	**Very Good** Fossil fuel displacement is significant in the near-term.	**Excellent** Land-based and aquatic production offer a large resource base. Organic residues, from many sources provide an early economic energy source.
Cost-Competitiveness with Fossil Fuels (Year 1990) Scale: Fair / Good / Very Good / Excellent	**Excellent** Some techniques require little additional cost, others require modest increases. Passive systems are very competitive with fossil fuels.	**Good** Water heating is competitive with electricity and oil in most regions of the U.S. Space heating is near-economic for northern regions. Space cooling is not economic.	**Good** Industrial process heat systems reducing use of electricity and oil can be competitive first in the Southwestern U.S. Solar electric systems augmenting existing oil-fired utility plants show early economic viability in the Southwest.	**Good** Competitive in remote electric markets. Prices need to decline by a factor of about 10 for grid-connected residential applications.	**Excellent** Competitive with fossil fuels, but assurance of long-term supply is a concern.	**Good** Ethyl alcohol from grains is now used in gasohol, which is competitive as a motor fuel. Anaerobic digestion of animal wastes is economic for rural uses. Gasifiers are near economic. Various liquefacation processes need cost reduction.
Degree of Social Acceptability Scale: Poor / Fair / Good / Very Good / Excellent	**Excellent** With appropriate designs, passive system components can be integrated aesthetically into a building.	**Excellent** Most systems can be aesthetically integrated into building structure.	**Good** Rural land areas covered by large collector fields are expected to be acceptable in selected arid areas in the Southwest. Areas adjacent to urban and scenic areas are expected to be very limited.	**Excellent** Systems produce electricity, which is well understood and widely accepted by consumers. There is little noise for home installation and system reliability is well established.	**Excellent** Aesthetically pleasing in home-heating; accepted practice in industry.	**Excellent** Gasohol is being accepted readily. Use of wastes for energy is appealing while reducing environmental impacts. Use of surplus farm products for energy is generally being endorsed. Increasing use of marginal crop lands is attractive.

continued ⟶

	Solar Collectors: Low-Temperature Passive	Solar Collectors: Low-Temperature Active	Solar Collectors: Intermediate- and High-Temperature	Photovoltaics	Biomass — Combustion	Biomass — Conversion
Environmental Acceptability Scale: Poor / Good / Very Good / Excellent	**Excellent** There are some potential concerns with indoor air pollution arising from close control of building air exchange.	**Excellent** There are minor problems in handling some waste products.	**Good** Some minor concerns exist regarding potential harm to desert environment.	**Excellent** Some minor concerns in manufacture. Generally, a very clean approach to power generation.	**Very Good** Particulate emissions and carbon dioxide may pose a problem.	**Good** Issues of land use, depletion of soil, erosion, and air pollution are of concern.
Institutional Barriers to U.S. Commercialization Scale: None / Few / Moderate / Many	**Few** Some resistance to modifying building codes; solar access legislation may be required.	**Few** Some modifications needed in building codes and solar access legislation. Many barriers have already been substantially resolved.	**Few** Regulations, land use and use of large reflective surfaces may pose some barriers.	**Few** Cost is a severe problem for other than remote electric applications. In addition, use in residential applications requires resolution of fire and safety building codes.	**Few** An industry structure needs to be developed for the collection and transportation of biomass.	**Few** There are persistent questions of competition between food and energy uses of biomass. Large land commitments for energy farming are required. Questions are being raised about net energy savings in production and use of biomass crops for fuels.
Availability of Financing in the United States Scale: Little / Fair / Good / Very Good / Excellent	**Excellent** Private investment is available as costs are small.	**Good** Substantial tax credits (40%) are available. First-cost financing is a continuing problem. Financing of domestic water heating systems is most easily obtained.	**Fair** Large capital investment and long payback hampers financing for process heat. Utilities require demonstration of reliability.	**Fair** Remote electric applications are economical with Federal and state tax credits. Large price reductions are required for widespread use in residential and utility applications.	**Excellent** Financing available where resources are adequate.	**Fair** Some Federal investment credit and loan guarantees are available but technology demonstrations are needed. Tax credits for homeowners and small system users are available.
Availability of Information in the United States Scale: Little / Fair / Good / Very Good / Excellent	**Good** Data on optimum designs are not readily available. Information on general approaches and system performance is available. Consumers and builders are not well informed about the potential benefits.	**Good** Detailed information on lifetime, reliability, costs and design details is available. Consumers are not well informed about selection, installation, and maintenance of systems.	**Fair** Parabolic trough data on performance and cost are available. Other systems need pilot system demonstration.	**Good** Lifetime, reliability, and cost data are available from domestic and international applications. Consumers are not well informed about potential uses.	**Very Good** Data are available through state and local organizations.	**Fair** Need further cost and performance data validation. Consumers need more education in the potential of these systems.

continued →

NEW AND RENEWABLE TECHNOLOGY SUMMARY-IN-BRIEF continued

	Wind	Ocean	Hydropower	Geothermal	Oil Shale	Tar Sands
Present State of Development Scale: R&D State Technically Ready Demonstration Stage Commercially Ready	**Demonstration Stage** Large systems being demonstrated at several sites; small systems are commercial.	**Technically Ready** Mini-OTEC and OTEC-1 experiments have been successfully deployed on the open ocean. Technology is now ready for a pilot plant development project.	**Commercially Ready** Mature technology. Small hydro needs some engineering and industrial development.	**Commercially Ready** About 800 MWe of geothermal power from dry steam wells in the U.S. Hotwater power plants are in the demonstration stage. Dry-hot-rock plants are in the R&D stage.	**Demonstration Stage** Engineering problems must be solved for modified in situ processes and above ground retorting.	**R&D Stage** Tar sands deposits in the U.S. are deeper than in Canada, may require in situ extraction.
Resource/Raw Materials Availability Scale: Fair Good Very Good Excellent	**Excellent** All materials and required resources are available.	**Good** Titanium production capacity expansion may be necessary; use of alternative materials may overcome this problem.	**Very Good** Construction material available locally; equipment requires lead time of one year or longer.	**Fair** Equipment requires special corrosion-resistant materials.	**Fair** Found in isolated areas where there may be significant water problems.	**Fair** Water availability may be a problem.
Fossil Fuel Displacement Potential (Year 2000) Scale: Fair Good Very Good Excellent	**Good** Wind resource regional and site-specific. Good potential for displacement in regions with most prevalent need, e.g., Northeast.	**Small** Commercial deployment by the early 1990s will be in tropical islands; but overall displacement is relatively small.	**Good** Expansion of existing large hydropower is limited, but use of small hydro could result in significant displacement.	**Fair** Although large in some areas, not highly significant from U.S. perspective, lack of experience in optimum rate of utilization.	**Good** Very large deposits, not uniformly distributed; meets transport fuel needs.	**Fair** U.S. deposits are nominal; removal of oil and raw material handling is difficult.
Cost-Competitiveness With Fossil Fuels (Year 1990) Scale: Fair Good Very Good Excellent	**Very Good** Small machines are economical in farm and rural applications. Utility applications will become competitive by 1990.	**Fair** Cost competitive in early markets that are oil dependent; power for U.S. mainland may be competitive in the late 1990s.	**Excellent** Extremely cost-effective in view of increasing fuel costs; marginal sites now economical.	**Good** Competitive where resource is abundant and high quality.	**Good** Costs are expected to be competitive when commercial plants begin operation.	**Fair** No data on economic feasibility of in situ operation.
Degree of Social Acceptability Scale: Poor Fair Good Very Good Excellent	**Very Good** Locating wind machines away from urban areas will mitigate problems of noise and television interference.	**Good** Off-shore operation presents no problems for urban areas; enhanced employment for fishing industry.	**Good** Precedents are well established; social issues are known and understood.	**Fair** Noise, brine and effluent disposal, and odor of hydrogen sulfide may not be acceptable if not controlled.	**Poor to Fair** Boomtown effect on small communities, large influx of workers and capital.	**Fair** Regional and local concerns similar to oil shale.

continued →

ENERGY GRAPHICS
64 Washburn Ave., Wellesley, MA 02181

126

NEW AND RENEWABLE TECHNOLOGY SUMMARY-IN-BRIEF continued

	Wind	Ocean	Hydropower	Geothermal	Oil Shale	Tar Sands
Environmental Acceptability Scale: Poor Fair Good Very Good Excellent	**Excellent** Minimal impact on air and water quality, and wildlife.	**Excellent** Some minor issues raised about possible changes in ocean ecology.	**Very Good** Some concerns of plant and animal life disruption, risk of dam failure.	**Good** Environmental impacts can be controlled.	**Poor to Fair** Air and water pollution problems; disruption of scenic land; disposal of waste materials.	**Fair** Environmental problems due to contamination of ground water by infiltration.
Institutional Barriers to U.S. Commercialization Scale: None Few Moderate Many	**Few** Federal legislation requiring utilities to sell and buy back power at equitable rates from residential and commercial users who have installed acceptable solar systems. Aesthetic considerations in urban areas may restrict commercialization.	**Moderate** Legal issues exist regarding legal jurisdiction, insurance, and operation of OTEC plants.	**Few** Hydropower is closely regulated; regulatory burdens are being eased for small hydro.	**Moderate** Will require environmental review for commercial plants; utility rate structure may be a problem for hydrothermal.	**Many** Lengthy environmental and societal impact reviews by Federal, state and local governments.	**Some** Large projects result in lengthy environmental and social impact reviews.
Availability of Financing in the United States Scale: Little Fair Good Very Good Excellent	**Very Good** Access to capital should not be difficult due to cost-effectiveness.	**Little** Large capital requirement and risks require demonstration of pilot plant, some government assistance may be available for early units.	**Very Good** Capital available for private and public projects; small hydro special investment tax credits are available.	**Good** Dependable and uninterruptible source can be developed in stages.	**Fair** Limited investment due to technological and social acceptance; loan guarantees available.	**Fair** Private investment will require demonstration; some Federal loan guarantees may be available.
Availability of Information in the United States Scale: Little Fair Good Very Good Excellent	**Very Good** Performance data on commercially available machines are being accumulated now. Data on large utility-scale prototypes are being obtained from current tests.	**Fair** Cost, performance, and resource data are available from early experiments, but pilot plant operational data needed.	**Excellent** Design construction procedures and costs are well documented and readily available.	**Very Good** Resource and technology information is available and is being distributed.	**Fair** Industry work remains proprietary; incomplete technical and economic data.	**Little** Data availability is limited due to deeper deposit and varying composition between sites.

Source: U.S. Department of Energy. 1981.

GLOSSARY

Active solar system. Any solar system that needs mechanical means such as motors, pumps, or valves to operate.

Agricultural and Industrial Process Heat (AIPH). Both an end-use application for thermal energy and the name applied to that portion of the U.S. solar program developing solar technologies for agriculture and industry (see IPH).

Anaerobic digestion. The process of using microorganisms that live only in the absence of oxygen to decompose organic materials into more valuable or more easily disposed of components. The chief use of anaerobic digestion is reduction of the volume of sewage sludge. It can also be used for the production of synthetic natural gas from organic wastes or from algae.

Array. (As in thermal or electrical solar array.) Several solar-collection devices interconnected in groups and in patterns.

Baseload plant (Baseload electricity). An electrical generation facility that is designed primarily to satisfy a continuous demand. Generally, capacity factors of baseload plants range from 60 percent to 90 percent.

Biofouling. The growth of small marine organisms that can decrease the heat transfer efficiency of heat exchangers used in seawater in ocean thermal energy conversion systems.

Biomass. Defined in the Energy Security Act (PL96-294) as "any organic matter which is available on a renewable basis, including agricultural crops and agricultural wastes and residues, wood and wood wastes and residues, animal wastes, municipal wastes, and aquatic plants."

British thermal unit (Btu). The amount of heat required to raise the temperature of one pound of water one degree Fahrenheit under stated conditions of pressure and temperature. (The Btu is equal to 252 calories, 778 foot-pounds, 1055 joules, and 0.293 watt-hours.) It is a standard unit in the United States for measuring quantity of heat energy.

Capacity factor. The ratio of the amount of electricity produced by a plant or system to its maximum theoretical productive capacity.

Capital-intensive. Requiring heavy capital investment. The energy industry, for example, is said to be capital-intensive rather than labor-intensive.

Closed-cycle ocean thermal energy conversion. A form of ocean thermal energy conversion (OTEC) using a working fluid (such as ammonia or propane) that can be vaporized by warm ocean surface waters and condensed with cool ocean waters, and recycled continuously in a closed loop.

Collector efficiency. The fraction of incoming solar radiation captured by the collector. For example, if the system captures half of the incoming radiation, the system is said to be 50 percent efficient. Efficiency is the capability of a collector to capture heat under various climatic conditions. There is no way a collector can be 100 percent efficient, that is, capture all the heat that falls on the collector; 55 percent is good under desirable weather conditions.

Concentrating collectors. Solar collectors designed to focus large amounts of solar radiation on a relatively small collection area to produce higher temperatures than those attainable by flat-plate collectors.

Conduction. The transfer of heat through matter by the movement of kinetic energy from particle to particle rather than by a flow of heated material; it also defines the way in which electricity travels through a wire or the way heat moves from a warm body to a cool one when the two bodies are placed in contact.

Convection. The transfer of heat by means of the upward motion of the particles of a liquid or a gas that is heated from beneath; transmission of heat by moving masses of air.

Conversion efficiency. The actual net output provided by a conversion device divided by the gross input required to produce the output.

Conversion system. A device or process that converts a raw energy form into another, more useful form of energy. Examples: conversion of wood into methanol or of sunlight into electricity.

Decentralized systems. In solar systems, complete, independent units of energy production to serve households, neighborhoods, villages, or factories, which are not dependent on other energy delivery grids.

Demand. The rate at which energy is delivered to or by a system expressed in kilowatts, kilovolt-amperes, or other suitable units, at a given instant or averaged over any designated period of time. The amount of energy required to satisfy the needs of a stated sector of the economy.

Department of Energy (DOE). In October 1977, the Department of Energy (DOE) was created to consolidate the multitude of energy-oriented government programs and agencies. The Federal Energy Administration and the Energy Research and Development Administration are now part of DOE. The Department is carrying out the National Energy Plan policy through a unified organization that coordinates and manages energy conservation, supply development, information collection and analysis, regulation, research, development, and demonstration. An Energy Information Administration within the Department organizes and analyze information so that it can be used by governments, industry, and the public.

Department of Housing and Urban Development (HUD). A U.S. Federal agency; responsible for most residential solar heating and cooling demonstration programs.

Diffuse insolation. The scattered solar power density (watts per square meter) falling on a surface from the sky, and, in the case of an inclined surface, reflected onto it from the ground as well; it does not include direct insolation. (See total insolation.)

Digester. A device in which anaerobic bacteria decompose organic matter to produce methane and carbon dioxide.

Direct insolation. The solar power density (watts per square meter) from the sun falling on a surface.

Distributed system. Used synonymously with decentralized system. (See definition.)

Domestic Policy Review (DPR). A cabinet-level, multi-agency review of the national solar energy program, initiated by the President on May 3, 1978, with results announced June 20, 1979.

Efficiency. The efficiency of an energy conversion is the ratio of the useful work or energy output to the work or energy input. For electrical generators, it is that percentage of the total energy

continued ⟶

GLOSSARY continued

content of a power plant's fuel which is converted into electricity, the remaining energy being lost to the environment as heat.

Energy Extension Service. A Government program to encourage the use of energy conservation and alternative energy technologies at the state and local levels. The program focuses primarily on homeowners, small businesses, public institutions, and state and local governments. The Service is similar to the USDA extension service.

Energy farm. A concept involving the cultivation of rapidly growing trees or other selected plants for the purpose of providing biomass that can be used as a fuel or converted into other energy products.

Ethanol. Ethyl alcohol or grain alcohol, $C_2H \bullet OH$. It is the alcohol contained in intoxicating beverages. Ethanol can be produced from biomass by the conversion process called fermentation and can be used as a motor fuel.

Eutectic salts. A chemical that has the property of changing from a solid to a liquid while maintaining a constant temperature. The heat energy that caused the transformation is stored in the eutectic liquid until the liquid returns to solid form and gives up heat. In application, eutectic salts are used as thermal reservoirs to conserve and then to release solar heat.

Evacuated tube collector. A collector designed to keep heat loss at a minimum. It is manufactured from concentric glass tubes with a vacuum between the tubes. These collectors are highly efficient at relatively high (about 100°C) temperatures.

Exajoule (EJ). A standard metric unit of energy equal to 10^{18} joules. One exajoule equals 0.948 quads. (See quads and joule.)

Feedstock. Material used for its chemical properties, rather than their value as fuel, e.g., oil used to produce plastics and synthetic fabrics. Also raw material that can be converted to one or more end-products (methanol or synthetic natural gas, for example). Biomass is an energy feedstock.

Fermentation. The process of decomposition of carbohydrates with the evolution of carbon dioxide or the formation of acid, or both.

Fiscal year (FY). U.S. Government's 12-month financial year, from October through September of the following calendar year; for example, FY 1978 extends from October 1977 through September 1978.

Flat-plate collector. A device for gathering the sun's heat, consisting of a shallow (usually metal) box with a glass or transparent plastic lid, where either air or liquid is circulated through the cavity of the box. The sun's energy is absorbed on a "black" plate and the entire box is insulated on the bottom and edges.

Fresnel lens and mirror. Used in solar concentrator devices to focus sunlight on solar cells in converting solar energy into electricity. The lens or mirror consists of stepped surfaces, allowing a thin design.

Fuel saver. A solar device used solely to save fuel at conventional fossil fuel-burning facilities. The conventional systems provide the needed system reliability.

Gasohol. A blend of alcohol and gasoline, typically 10 percent agriculturally derived ethanol and 90 percent unleaded gasoline.

Gasifier. A vessel in which gasification takes place.

Geothermal energy. The heat energy available in the rocks, hot water, and steam in the earth's subsurface.

Gigawatt. Power unit equal to one billion (10^9) watts, one million kilowatts, or one thousand megawatts.

Greenhouse effect. The heating effect of the atmosphere upon the earth. Light waves from the sun pass through the air and are absorbed by the earth. The earth then reradiates this energy as heat waves that are absorbed by the carbon dioxide in the air which behaves like glass in a greenhouse, allowing the passage of light but not of heat. Many scientists theorize that an increase in the atmospheric concentration of carbon dioxide can eventually cause an increase in the earth's surface temperature.

Head. The differential of pressure causing flow in a fluid system, usually expressed in terms of the height of a liquid column that the pressure will support.

Heat exchanger. A device that transfers heat from one fluid (liquid or gas) to another, or to the environment.

Heat pump. A reversible heating and cooling mechanism that can produce additional usable heat from the amount stored, such as a mechanical refrigeration system which is used for air cooling in the summer and which, when the evaporator and condenser effects are reversed, can absorb heat from the outside air or water in the winter and raise it to a higher potential so that it can also be used for winter heating.

Heliostat. A device that contains a mirror moved by a control mechanism to reflect the light of the sun in a particular direction.

High-head. Refers to generation of hydropower using large dams (compare, low-head).

Hybrid solar system. Solar energy systems that combine both direct thermal (passive) and indirect solar design elements and equipment into one system; for example, solar collectors on a roof providing heat to the direct thermal storage component located inside the insulated shell of the structure.

Hydropower (Hydroelectricity). Power produced by falling water.

Hydrostorage. Use of a dam to store energy: water is pumped into a reservoir and released when power needs to be generated.

Industrial Process Heat (IPH). The end-use applications of thermal energy (at a wide range of temperatures) in industry. The term is also used to indicate that portion of the U.S. national solar program that deals with this end-use. (See AIPH.)

In situ. In the natural or original position or location. In situ production of oil shale, for instance, is an experimental technique in which a region of shale is drilled, fractured, and set on fire. The volatile gases burn off, the oil vaporizes, then condenses and collects at the bottom of the region from which it can be recovered by a well.

Insolation. The rate at which energy reaches the earth's surface from the sun. Usually measured in Btu per square foot per day or joules per square meter per day.

International Solar Energy Society (ISES). The principal, non-governmental worldwide organization for exchanging information on all forms of solar energy.

Joule (J). The meter-kilogram-second (mks) unit of work or energy. It equals 1 watt-second or 10^7 ergs, or approximately 0.7375 foot-pound, or 0.2390 gram-calorie.

Kilowatt (kW). A unit of power equal to 1,000 watts or to energy consumption at a rate of 1,000 joules per second. It is usually used for electrical power. An electric motor rated at one horsepower uses electrical energy at a rate of about 3/4 kilowatt.

Kilowatt-hour (kWh). The unit of energy equal to that expended in one hour at a rate of 1 kilowatt; or 3,413 Btu.

continued ⟶

GLOSSARY *continued*

Life-cycle cost. The total costs for the purchase, installation, operation, and maintenance of a system over its useful life. The accumulation generally includes a discounting of future costs to reflect the relative value of money over time.

Line-focusing collector. A mirror system, usually consisting of a parabolic trough, that concentrates the incident solar radiation on a line where it can be converted to useful energy. (See point-focusing collector.)

Low-head. Refers to generation of hydropower with relatively small dams (approximately 30 megawatts or less). (Compare, high-head hydropower.)

Megawatt (MW). A unit of power. A megawatt equals 1000 kilowatts, or one million watts.

Methane. A colorless, odorless, flammable, gaseous hydrocarbon that is a product of the decomposition of organic matter. It is used as a fuel and as a raw material in chemical synthesis. The chemical formula is CH_4. CH_4 is the prime constituent of natural gas. Methane is made by certain biomass conversion processes.

Methanol. A light, volatile, flammable, poisonous, liquid alcohol, CH_3OH, formed in the destructive distillation of wood or made synthetically and used especially as a fuel, a solvent, an antifreeze, or a denaturant for ethyl alcohol, and in the synthesis of other chemicals. Methanol can be produced as fuel for motor vehicles.

MBtu (often MMBtu). Million Btu's (British thermal units).

Module. The smallest, complete assembly of solar cells, optics, and other components designed to generate DC power.

Ocean thermal energy conversion (OTEC). Process of using the warm ocean water (ocean's sun-heated top layer) to vaporize a working fluid to run a turbine to produce electricity.

Off-peak service. The period during a day, week, month, or year when the load being delivered by an energy system is not at or near the maximum volume delivered by that system for the corresponding period of time.

Oil shale. A very fine-grained sedimentary rock that contains enough organic matter (hydrocarbon) to yield 10 gallons or more oil per ton when properly processed. Some shales yield much more oil, and in the United States some thin layers of shale have been reported to yield 140 gallons of oil per ton. Most shales that are of commercial interest yield from 25 to 65 gallons of oil per ton. Some foreign deposits that contain shales yielding between 10 and 25 gallons per ton have been mined on a large scale. Many other organic-rich shales yield less oil (1 to 10 gallons per ton), but these shales are so low grade that they usually are not called "oil shale."

Open-cycle ocean thermal energy conversion. (See closed-cycle ocean thermal energy conversion.) In open-cycle OTEC systems, the working fluid is ocean water that is vaporized.

Parabolic collector. A device for collecting the sun's energy that utilizes a bowl-shaped reflector. The reflector, which is mirrored, concentrates the radiation, producing temperatures as high as 1649°C.

Parabolic dish. A parabola-shaped device covered with a mirrored surface that focuses sunlight at a single point; used in solar concentrators with concentration ratios typically from 1,000 to 10,000.

Passive solar system. A system that uses gravity, heat flows, or evaporation to operate without mechanical devices to collect and transfer energy, such as sun-facing windows.

Payback period. A traditional measure of economic viability of investment projects. For energy projects, a pay-back period is defined in several ways—one of which is the number of years required to accumulate fuel savings equal to the capital cost of the system. (See life-cycle cost.)

Peak watt (W_p). A unit of measure used in rating the performance of photovoltaic panels and arrays. A panel rated at 1 peak watt will deliver 1 watt under specific standard noontime operating conditions (including a solar insolation of 1 kilowatt per square meter).

Phase change. The physical transformation of a substance from one state—solid, liquid, or gaseous—to another. Associated with such a change of state is a large absorption or release of energy known as the heat of fusion or vaporization.

Photosynthesis. The process in plants by which carbohydrates are compounded from carbon dioxide and water in the presence of sunlight and chlorophyll. Photosynthesis is the primary method of bioconversion of solar energy into forms more useful to society.

Photovoltaic cell. A type of semiconductor in which the absorption of light energy creates a separation of electrical charges. The separation creates an electrical potential. The net effect is direct conversion of light into electricity. Typical materials used in the construction of photovoltaic cells are silicon, cadmium sulfide, and gallium arsenide.

Photovoltaic process. Process of converting light rays directly into electricity without going through intermediate steps involving turbines and generators. In photovoltaics, solar cells convert sunlight into a stream of electrons easily converted into household current.

Point-focusing collector. Any mirror or system of mirrors or lenses that concentrates sunlight on a point.

Power tower. A tower placed so that the reflected direct radiation from heliostat mirrors can be focused onto a receiver at its top. Heat exchange takes place at the receiver on the tower, and the heated fluid is used in a conventional power system.

Primary energy. Energy in its naturally occurring form (coal, oil, uranium) before conversion to end-use forms.

Process heat. Heat that is used in agricultural and industrial operations.

Pyrolysis. Decomposition by the action of heat; thermal decomposition of organic compounds in the absence of oxygen.

Quad. One quadrillion (10^{15} or 1,000,000,000,000,000) Btu.

Rankine engine. A reversible heat engine.

Receiver. The solar thermal component of a solar concentrator system. A receiver is designed to operate under concentrated sunlight and to absorb thermal energy.

Renewable resources. Sources of energy that are regenerative or virtually inexhaustible, such as solar, wind, ocean, biomass and hydropower energy. Geothermal energy is sometimes also included in the term.

Retrofit. The installation of equipment subsequent to the completion of initial construction, such as the installation of solar collectors and other hardware onto an existing house to convert it for solar heating.

Roof pond. A passive solar collection system in which plastic bags of water on the roof are appropriately exposed or insulated during the day and night, depending on the thermal requirements of the building for summer and winter operation. Heat transfer (into or out of the bags) to the structure is directly through the ceiling.

continued ——————▶

GLOSSARY *continued*

Semiconductor. A crystal system in which, though the electrons are ionically bound, a slight rise in temperature frees the valence atoms so that the system becomes a conductor; an example is germanium.

Shale oil. A crude oil obtained from bituminous shales by submitting them to destructive distillation in special retorts.

Silviculture. The technology of raising trees, or forest management.

Solar cell. A device which converts radiant energy directly into electric energy by the photovoltiac process. Each cell produces a small potential difference, typically about 0.5 volts; an array of cells can provide a useful electric power capacity.

Solar collector. A device used to gather and accumulate the sun's energy or solar radiation. Nearly all non-concentrating collectors have a layer of glass on top (some have plastic) to trap the heat once it passes into the collector. The medium used to transfer the heat to the rest of the system varies. Another type of collector, the focusing collector, employs a concave mirror turned by a motor to follow the sun as the earth moves. The basic function of the solar collector is to capture the sun's heat.

Solar constant. The average amount of solar radiation reaching the earth's atmosphere per minute. The solar constant is measured on a plane perpendicular to the path of the radiation. Its value is 1.36 kilowatts per square meter (or 430 Btu per square foot per hour).

Solar energy. The energy transmitted from the sun in the form of electromagnetic radiation. Although the earth receives about one-half of one billionth of the total solar energy output, this amounts to about 420 trillion kilowatt-hours annually.

Solar Energy Research Institute. The Solar Energy Research, Development and Demonstration Act of 1974 called for the establishment of a Solar Energy Research Institute whose general mission would be to support the nation's solar energy program and foster the widespread use of all aspects of solar technology, including direct solar conversion (photovoltaics), solar heating and cooling, solar thermal power generation, wind energy conversion, ocean thermal conversion, and biomass conversion.

Solar pond. A pond with a blackened base containing (stratified) brackish water or insulation that is used to collect and store solar heat.

Solar Power Satellite (SPS). A proposed satellite that would be put into stationary orbit above the earth to convert solar energy to electricity and then to microwave (or possibly laser) energy, to be beamed to earth for conversion back to direct and then alternating current electricity.

Solar rights. Continued access to sunlight for installed solar systems.

Solar power thermal conversion systems. Thermal conversion systems involving an extensive array of pipes heated by the sun's rays. In one concept, nitrogen flowing through the pipes would gather the heat and transport it to molten salt. The molten salt can be heated to a temperature of about 538 °C for production of steam, which would power conventional turbines at a projected efficiency of about 30 percent. The area required to supply energy to a 1000-megawatt power plant would be about 10 square miles of collection surface; a 300,000-gallon reservoir of molten salt for energy storage would be necessary for nights and cloudy days.

Solar total energy system (STES). Solar total energy systems use the sun's heat to generate electricity and the leftover or residual heat for other purposes. Several technical approaches are being investigated as applicable to STES. These include small central receivers, distributed collectors, and photovoltaics.

South wall. See Trombe wall. This would be termed a "North wall" in the southern hemisphere.

Stirling hot-air engine. The Stirling engine can use solar energy or can burn any type of fuel to generate the heat used in the engine. Like the internal combustion engine, the Stirling has a cylinder and pistons. However, the solar energy is directed to the outer walls or the fuel is burned outside of the cylinder (external combustion). The heat is used to expand a gas in the cylinder, which pushes down on the piston to produce the power stroke. Problems that remain to be solved include the handling of the very large quantity of waste heat which must be removed by the radiator and the need for heater heads which can contain the high-temperature and high-pressure gas.

Synthetic natural gas (SNG). A gaseous fuel manufactured from coal or biomass, containing almost pure methane, CH_4, and produced by a number of gasification schemes. SNG contains 95 percent to 98 percent methane, and has an energy content of 980 Btu to 1035 Btu per standard cubic foot, about the same as that of natural gas.

Tar sands. Sand impregnated with petroleum that has turned to viscous or solid bitumen; hydrocarbon-bearing deposits distinguished from more conventional oil and gas reservoirs by the high viscosity of the hydrocarbon, which is not recoverable in its natural state through a well by ordinary oil production methods.

Tennessee Valley Authority (TVA). A large Federal electric power agency that is also a major purchaser of renewable energy systems and active in renewable energy experiments.

Terawatt (TW). Power unit equal to 10^{12} watts or one million megawatts.

Thermal conversion. The transformation of heat into other forms of energy such as electricity or shaft power. Heat may be supplied from the sun either directly (solar collector) or indirectly (ocean systems or biomass).

Thermal mass. Also thermal inertia. The tendency of a building or components with large quantities of heavy materials to change temperature very slowly; also, the overall heat-storage capacity of a building.

Thermosiphon. The principle that makes water circulate automatically between a solar collector and a storage tank above it, gradually increasing its temperature. A solar heating system that uses natural convection to transport heat from the collector to storage by appropriately locating the storage in relation to the collector.

Total energy system (TES). A packaged energy system of high efficiency, utilizing solar energy or gas-fired turbines or engines that produce electrical energy and utilize exhaust heat in applications such as heating and cooling. (See also solar total energy systems).

Total insolation. The sum of direct and diffuse insolation.

continued ——→

GLOSSARY continued

Trombe wall. A passive solar heating system that combines the solar collector and heat storage in one, sun-facing wall unit. The system consists of a thick concrete wall painted black on its outer face. Sheets of glass are placed in front of this wall with an airspace between. Air from the rooms of the building passes through openings at the foot of the wall and enters the airspace where it is heated by the sun. As it warms, the air rises up the air cavity by natural convection, and passes back into the building interior again through a second series of openings at the top of the wall. In order to arrest the flow of warm air into the building in summer, the openings in the wall are blocked by shutters.

Wind energy conversion system (WECS). A wide variety of machines have been designed to convert wind energy into useful power. The large-scale systems currently being developed most often utilize a horizontal axis design: any design will require siting in high average wind speed areas.

Source: U.S. Department of Energy. 1981.

ACRONYMS

AID	Agency for International Development, the principal U.S. bilateral development assistance agency
AIPH	Agricultural and Industrial Process Heat
bbl	Barrels (of oil). One barrel equals 42 American gallons, 306 pounds, 5.6 cubic feet. The heat content is approximately 5.8×10^6 Btu/bbl.
Btu	British thermal unit (See Glossary)
DOE	Department of Energy (See Glossary)
DPR	Domestic Policy Review (See Glossary)
EJ	Exajoule, 10^{18} or one billion billion joules (See Glossary)
EPA	Environmental Protection Agency
FERC	Federal Energy Regulatory Commission
FY	Fiscal Year (See Glossary)
GJ	Gigajoule, 10^9 or one billion joules
GNP	Gross National Product
IPH	Industrial Process Heat (See Glossary)
ISES	International Solar Energy Society (See Glossary)
kW	Kilowatt (See Glossary)
kWh	Kilowatt-hour (See Glossary)
MBtu	Million (10^6) British thermal units (See Glossary)
MJ	Megajoule, 10^6 or one million joules
MW	Megawatt, a unit of power equal to 10^6 watts
MWh	Megawatt-hour, a unit of energy, equal to 10^6 watt-hours
NASA	National Aeronautics and Space Administration, active in several solar technology areas, has the major U.S. responsibility for the SPS (Solar Power Satellite)
NSF	National Science Foundation, the lead Federal agency for solar energy research and development until 1975
OTEC	Ocean Thermal Energy Conversion (See Glossary)
PURPA	Public Utilities Regulatory Policy Act
R&D	Research and Development
RD&D	Research, Development and Demonstration
SERI	Solar Energy Research Institute (See Glossary)
SHAC	Solar Heating and Cooling (See Glossary)
SPS	Solar Power Satellite or Space Power Systems; also sometimes used to refer to Small Power Systems (See Glossary)
STES	Solar Total Energy System (See Glossary)
TES	Total Energy System (See Glossary)
TVA	Tennessee Valley Authority (See Glossary)
TW	Terawatt (See Glossary)
USDA	United States Department of Agriculture, often termed DOA (responsible for several parts of the national solar program)
WECS	Wind Energy Conversion Systems (See Glossary)
W_p	Peak watt (See Glossary)

☆ U.S. GOVERNMENT PRINTING OFFICE: 1981—341-065/165

Source: U.S. Department of Energy. 1981.

VII. TABLES

Units of Measure, Conversion Factors, Price Deflators, and Energy Equivalents

Weight

1 short ton	contains	2,000 pounds
1 metric ton	contains	1.102 short tons
1 long ton	contains	1.120 short tons

Conversion Factors for Crude Oil (Average Gravity)

1 U.S. barrel	contains	42 U.S. gallons
1 U.S. barrel	weighs	0.136 metric tons
1 metric ton	contains	7.33 barrels
1 short ton	contains	6.65 barrels

Conversion Factors for Uranium

1 short ton (U_3O_8)	contains	0.769 metric tons
1 short ton (UF_6)	contains	0.613 metric tons
1 metric ton (UF_6)	contains	0.676 metric tons

Approximate Heat Content of Refined Petroleum Products (Million Btu per Barrel)

Asphalt	6.636
Aviation gasoline	5.048
Butane	4.326
Butane-propane mixture*	4.130
Distillate fuel oil	5.825
Ethane	3.082
Ethane-propane mixture**	3.308
Isobutane	3.974
Jet fuel—kerosene type	5.670
Jet fuel—naphtha type	5.355
Kerosene	5.670
Lubricants	6.065
Motor gasoline	5.253
Natural gasoline	4.620
Petrochemical feedstocks	
Naphtha 400° F	5.248
Other oils over 400° F	5.825
Still gas	6.000
Petroleum coke	6.024
Plant condensate	5.418
Propane	3.836
Residual fuel oil	6.287
Road oil	6.636
Special naphtha	5.248
Still gas	6.000
Unfinished oils	5.825
Wax	5.537
Miscellaneous	5.796

*60 percent butane and 40 percent propane.
**70 percent ethane and 30 percent propane.

Gross National Product (GNP) Implicit Price Deflators. 1972 = 100.

1950	53.56		1966	76.76
1951	57.09		1967	79.06
1952	57.92		1968	82.54
1953	58.82		1969	86.79
1954	59.55		1970	91.45
1955	60.84		1971	96.01
1956	62.79		1972	100.00
1957	64.93		1973	105.69
1958	66.04		1974	114.92
1959	67.60		1975	125.56
1960	68.70		1976	132.11
1961	69.33		1977	139.83
1962	70.61		1978	150.05
1963	71.67		1979	162.77
1964	72.77		1980	177.40
1965	74.36			

Note: All GNP Implicit Price Deflators were revised in late 1980 and early 1981.
Source: U.S. Department of Commerce, Bureau of Economic Analysis.

Electricity Consumption . . . . 3,412 Btu/kilowatt-hour

Hardwood (Dry) . . . . 8,000-9,000 Btu/pound

Using Thermal Equivalent Conversion Factors

The various energy sources are converted from the original units (i.e., short tons, cubic feet, barrels, and kilowatt-hours) to the thermal equivalent using British thermal units (Btu). A Btu is the amount of energy required to raise the temperature of 1 pound of water 1 degree Fahrenheit at or near 39.2 degrees Fahrenheit. One Btu is equivalent to about 252 IT (International Steam Table) calories.

Btu conversion factors for hydrocarbon mixes are the weighted average of the Btu content of all hydrocarbons included in the mix. All Btu factors are computed from *final* annual data. If the current year's final data are not available, Btu conversion factors for the latest final annual data are used.

Source: U.S. Department of Energy. 1981.

Energy Equivalents

One million Btu equals approximately:

89	pounds of bituminous coal and lignite production (1980)
125	pounds of oven-dried wood
8	gallons of motor gasoline or enough to move the average passenger car about 110 miles
10	therms of natural gas (dry)
11	gallons of propane
0.9	average daily per capita energy consumption in the United States (1980 rate)
2	months of dietary intake of a laborer
20	cases (240 bottles) of table wine

One million Btu of fossil fuels burned at electric utilities can generate about 100 kilowatt-hours of electricity, while about 300 kilowatt-hours of electricity generated at electric utilities can produce about one million Btu of heat.

One quadrillion Btu equals approximately:

44	million short tons of bituminous coal and lignite production
63	million short tons of oven-dried wood
1	trillion cubic feet of natural gas (dry)
170	million barrels of crude oil
500	thousand barrels per day of crude oil for one year
25	days of petroleum imports into the United States (1980 rate)
29	days of motor gasoline consumed in the United States (1980 rate)

One barrel of crude oil equals approximately:

5.7	thousand cubic feet of natural gas (dry)
0.26	short tons of bituminous coal and lignite production
1,700	kilowatt-hours of electricity consumed

One short ton of bituminous coal and lignite production equals approximately:

3.9	barrels of crude oil
22	thousand cubic feet of natural gas (dry)
6,600	kilowatt-hours of electricity consumed

One thousand cubic feet of natural gas equals approximately:

0.18	barrels, or 7.5 gallons of crude oil
0.045	short tons (or 90 pounds) of bituminous coal and lignite production
300	kilowatt-hours of electricity consumed

One thousand kilowatt-hours of electricity equal approximately:

0.59	barrels of crude oil (although it takes about 1.7 barrels of oil to produce 1,000 kWh)
0.15	short tons of bituminous coal and lignite production (although it takes about 0.5 short tons to produce 1,000 kWh)
3,300	cubic feet of natural gas—dry (although it takes about 10,000 cubic feet to produce 1,000 kWh)

Approximate U.S. Daily Per Capita Consumption of Types of Energy in 1980

Natural gas (dry), including consumption at electric utility powerplants	250 cubic feet
Natural gas (dry), excluding consumption at electric utility powerplants	200 cubic feet
Coal, including consumption at electric utility powerplants	17 pounds
Coal, excluding consumption at electric utility powerplants	3.4 pounds
Hydropower electricity	3.7 kilowatt-hours
Nuclear power electricity	3.1 kilowatt-hours
Electricity, including hydropower and nuclear power electricity	26 kilowatt-hours
Refined petroleum products, including consumption at electric utility powerplants	3.2 gallons
Refined petroleum products, excluding consumption at electric utility powerplants	3.0 gallons
Motor gasoline	1.2 gallons

Source: U.S. Department of Energy. 1981.